AIDE TO MILITARY INSTRUCTION

WITH

PLANS AND DIAGRAMS

BY

L. De T. PREVOST, F.G.S.

MAJOR ARGYLL AND SUTHERLAND HIGHLANDERS;
FORMERLY
BRIGADE-MAJOR OF INFANTRY AT ALDERSHOT

The Naval & Military Press Ltd

published in association with

Published by
The Naval & Military Press Ltd
Unit 10 Ridgewood Industrial Park,
Uckfield, East Sussex,
TN22 5QE England
Tel: +44 (0) 1825 749494
Fax: +44 (0) 1825 765701
www.naval-military-press.com

in association with

ROYAL
ARMOURIES

In reprinting in facsimile from the original, any imperfections are inevitably reproduced and the quality may fall short of modern type and cartographic standards.

PREFACE.

THIS little work makes no pretension to being an exhaustive treatise. It consists rather of a series of notes strung together for purposes of instruction, originating in General Order 50 of last year, which introduced the training of men under their own officers.

The object in view has been to make the contents of practical value chiefly to young officers; who are recommended, however, to seek further information on the various subjects in standard works.

As it is somewhat difficult to determine where the knowledge required of the soldier shall end and that of his superiors commence, it has been deemed advisable to mark with an asterisk such portions of the text as more directly apply to the non-commissioned officers and men.

MARYHILL BARRACKS, GLASGOW,
 January, 1884.

ERRATA.

Page 17 (4) For "pealed" read "peeled."

Diagram, facing page 44, figure 33, for "Quater Column" read "Quarter Column."

BOOKS CONSULTED.

The following works have been consulted:—

The Soldier's Pocket-Book. General Lord Wolseley, G.C.B., &c., &c.

Operations of War. Lieut.-General Sir E. B. Hamley, K.C.B.

Précis of Modern Tactics. The late Colonel Home, R.E.

Franco-German War, 1870–71. *Official Account.*

Tactical Deductions from the War, 1870–71. Lieut.-Colonel A. V. Boguslawski.

Les Manœuvres de l'Infanterie, 1881.

Defence and Attack of Positions. Colonel Schaw, R.E.

Field Artillery. Major Pratt, R.A.

What to Observe and How to Report It. Colonel Hale (H.P.), R.E.

Journals of the United Service Institution.

CONTENTS.

	PAGE
PREFACE	v
LIST OF WORKS CONSULTED	vii
LIST OF PLATES	xi
INTRODUCTION	xiii

CHAPTER I.
ELEMENTARY TACTICS 1

CHAPTER II.
ENCAMPMENTS, FIELD COOKING, SHELTER TRENCHES 8

CHAPTER III.
THE THREE ARMS 24

CHAPTER IV.
INFANTRY IN ATTACK AND UNDER OTHER CIRCUMSTANCES 48

CHAPTER V.
ADVANCE GUARDS, FLANKING PARTIES, REAR GUARDS 69

CHAPTER VI.
OUTPOSTS 84

CHAPTER VII.
MARCHES 110

CHAPTER VIII.
MINOR OPERATIONS 126

CHAPTER IX.
OBSTACLES, AND OTHER ACCESSORIES TO DEFENCE 161

CHAPTER X.
WORKING PARTIES, ESCALADE, HASTY DEMOLITIONS 180

CHAPTER XI.
CROSSING RIVERS, AND BRIDGING 198

CHAPTER XII.
RECONNAISSANCE AND FIELD SKETCHING 220

INDEX 251

LIST OF PLATES.

Plate	I. Natural Cover	To face page	1
,,	II. Camp of an Infantry Battalion	,,	9
,,	III. Artificial Cover	,,	18
,,	IV. Ideal Disposition of Outposts	,,	84
,,	V. Disposition of a Brigade on Outpost Duty	,,	105
,,	VI. Order of March of a Division	,,	121
,,	VII. Defence of Villages	,,	139
,,	VIII. Defence of a Wood	,,	146
,,	IX. Passage of the River Douro	,,	158
,,	X. Conventional Signs	,,	244

Figures 1 to 136 Facing Pages they Illustrate.

INTRODUCTION.

IN the present condition of warfare the army requires much careful training, high discipline, and steady practice, if we are to hold our own with the troops of other nations. In foreign armies short service obtains, necessitating more time and care in perfecting the material during the period of service with the colours. We now, having adopted a similar system, must follow in the same direction, bearing in mind the greater amount the soldier has to learn in a shorter time than formerly in order to become valuable in the field, and that instruction limited to regimental routine within the walls of a barrack square is insufficient for this purpose.

The recent general order on the subject of military instruction is a decided step in the right direction. Although broad principles are necessary as a basis, a company officer should be allowed every latitude in carrying out the details, by which he will have a greater incentive to study his profession, feeling that the responsibility rests upon himself of making his non-commissioned officers and men efficient.

System of training.

In order that the training bring forth good results, it ought to be as varied and as interesting as circumstances will permit. The instructor should make himself thoroughly conversant beforehand with the subject selected for each day; and it has been found by experience that men of little or no education will evince an intelligent interest in the work, if it is first patiently explained to them in clear, simple language, with the assistance of a blackboard and a few diagrams, and afterwards executed practically on the ground.

In the education of the soldier there are three elements: drill, field-training, and discipline.

Drill.

Drill published by authority must not be deviated from. By it men are taught to move with precision and alacrity at word of command. Here discipline first comes into play, laying the foundation for higher training; and there is danger in relaxing it—for if

Discipline.

men are unable to maintain distances, intervals, and direction on a level parade, they can scarcely be expected to do so over broken ground during manœuvres. Success in tactics depends upon the proper application of drill; and until a battalion is pliable, and easily handled, it cannot be considered fit to take the field.

Field training.

The second element includes manœuvres and minor operations, which should be as faithful representations as possible of what really would occur in war; and also all duties incident on active service, which subjects are treated in the following pages.

Musketry.

The time set apart for the annual course of musketry must be entirely devoted to this most important branch of training, and no pains should be spared in perfecting the men's shooting by careful supervision, encouraging emulation, and organising shooting matches.

But the experiences of Majuba Hill, and the great expenditure of ammunition in Egypt without proportionate results, point to the necessity of attaining physical superiority as well as proficiency in using the rifle. The German infantry soldier in 1870-71 was armed with an infinitely inferior weapon to the Frenchman's. His supremacy lay rather in the individual man himself than in handling his rifle.

For the preliminary instruction of the recruit, the following distribution of time has been found to answer well :—

<small>Preliminary instruction of the recruit.</small>

	In Summer.	In Winter.
Drill,	6.30 to 7.30 A.M.	8.15 to 9.15 A.M.
Gymnasium,	9 ,, 10 ,,	9.45 ,, 10.45 ,,
Drill,	11 ,, 12 noon.	11.30 ,, 12.30 P.M.
,,	2 ,, 3 P M.	2 ,, 3 ,,
School,	3.30 ,, 5 ,,	3.30 ,, 5 ,,

After two months he will have acquired sufficient knowledge of marching, drill, and rifle exercises to be prepared for a course of musketry instruction, which, in moderate weather, can be easily finished in three weeks, and at the end of the first two months he should also have been dismissed the recruits' course at the gymnasium. Musketry concluded, he should revert to drill, and be exercised in the more advanced stages, e.g., bayonet exercise in quick time, company drill, skirmishing, the elements of the attack, and duties on guard and sentry, which will occupy five weeks. Drill being substituted for gymnasium, he will then have four hours daily.

Interior economy. * Concurrently with the above, the recruit must be instructed in the interior economy of the barrack-room. The first thing to teach a lad on joining is how to dress, clean his arms and accoutrements, make up his cot, arrange his kit on the shelf, and lay it out for inspection, in strict conformity with standing orders. The duty of this instruction devolves on the senior N. C. officer of each room, who must distribute the recruits among the older soldiers; and the latter are required to show an example of cleanliness and regularity. No man is to mount guard or piquet, or leave the room for his daily employment, without arranging his cot and kit in proper order, and cleaning his arms. Without insisting on needless minutiæ, this system should exist throughout the week, and not only for periodical inspections.

Running drill. Running drill is most essential, and should be practised both during the recruits' course and frequently at other times; otherwise young soldiers will be found unfitted for a day's march, as many of them take but little exercise outside the barrack square.

Such, then, is an outline of the preliminary instruction, which will extend over sixteen weeks, or four months, and at the expiration of this time, under ordinary circumstances, the recruit will be able to take his place in the ranks as a duty soldier, ready to receive the further training, which now requires our attention.

AIDE TO MILITARY INSTRUCTION.

CHAPTER I.

ELEMENTARY TACTICS.

THERE are a few definitions within the comprehension of every one, and should be learnt in order to understand something of the art of war. Definitions.

The expression "tactics" must be distinguished from "strategy." "Tactics" relates to the battlefield; it means the art of forming troops in "order of battle" and effecting changes in their dispositions according to the progress of events; * Tactics.

Whereas "strategy" has reference to the whole campaign, and all preliminary arrangements for engaging the enemy. Strategy.

By the "base of operations" is meant the frontier, coast line, town, or district, upon which an army relies for supplies and support during a campaign. Base.

The "line of operations" is that by which an army advances from its base to meet the enemy. Line of operations.

Line of communications.	The "line of communications" is that by which an army receives reinforcements, supplies, and ammunition from the base, and sends back the sick and wounded to the rear.
Theatre of war.	The "theatre of war" is the whole area of country, in any part of which the contending armies can operate and come into collision with each other.
Strategical point.	A "strategical point" is one which will strengthen a line of operations. A "tactical point" is one which on the field of battle will facilitate attack, or impede the enemy's advance.
Moral agents.	There are certain moral agents in war, viz.:—the character and skill of commanders; the elation, emulation, or depression of soldiers; stratagems; means of obtaining and distributing information. It was a saying of Napoleon that the "moral was to the physical force as 3 is to 1."

This great general laid down various maxims of war, of which the chief are :—

Some maxims.

To place the several parts of an army in such positions that they may be able to support each other.

To be able to concentrate easily.

To be able to collect a larger force at a given point than the enemy can oppose to it.

To operate on the enemy's communications without exposing one's own.

Never to abandon a line of communication from over confidence.

Always to operate on interior lines.

The limit of random infantry fire may be taken as 1,400 yards; that of aimed infantry fire at 800 yards.

<small>* Three zones of fire.</small>

There may be considered three zones of fire action, viz. :—

i. Between 2,500 and 1,600 yards from an enemy's position. Good practice can be made with shell at 2,500 yards. We have only to fear artillery in this zone. This arm will chiefly direct its fire against opposing batteries. It has the chief rôle of action, and considerable choice of ground.

ii. Between 1,600 and 700 yards from the position, when artillery fire is very destructive, and unaimed infantry fire comes into play. Both artillery and infantry fight together in this zone, the chief object of the former being to aid the advance of the infantry.

iii. From 700 yards onward.—Here infantry fires against infantry, and both these arms fire against the enemy's guns. Artillery find it difficult to change position owing to infantry fire.

Troops should be so arranged for attack as to develop the most destructive fire compatible with facility of movement, and at the same time to offer the least possible target to the enemy.

<small>* General arrangements for attack.</small>

Artillery, concentrated in front, prepares the way for the other arms, and the general formation of the force will greatly depend on the nature of the ground. It will consist of a first and second line, and a reserve, with cavalry and artillery on the flanks, and between the intervals of the infantry, as the requirements of the attack admit.

Necessity for deployment. * Infantry arrives on the field of battle in column of route, or in mass, from which they are deployed into these three lines of columns at deploying interval. This arrangement allows the other arms to pass through, and subsequent movements are more easily adapted to the ground. It is evident that the deployment should be effected, when opposed to artillery, at from 2,000 to 2,500 yards from the position.

The attacking force should be sufficiently supported in the event of being successful; or if it is repulsed, the retreat should be covered. The clearest instructions should be given to the several officers as to the object of the attack and the movements intended, so that the entire force may be utilised.

Proportion of the three arms. The proportion of the three arms to each other on active service is:—Artillery, about 3 guns per 1,000 infantry; cavalry, about one fourth the infantry. The tactical units are the battalion, the battery, the squadron. 3 battalions form an infantry brigade.

The division as a unit. A division is the first important combined unit in an army; it is composed of the three arms, and is sufficiently strong, say from 10,000 to 14,000, to be capable of independent action. It is about the largest number that one general officer can properly superintend.

Our division is smaller than those of continental armies. Its organisation is given in Part V. of the *Field Exercises*. When the division has no tents its composition is as under:—

Staff	55 officers and men.	Personnel of a division.
2 infantry brigades (6 battalions)	6,706	
1 battalion rifles	1,097	
1 regiment cavalry	653	
3 batteries of artillery (1 9-pounder, 2 16-pounders)	573	
1 field company, R.E.	202	
Infantry and artillery, ammunition, reserve columns	214	
Supply and transport in addition to that possessed by brigades	189	
Medical, veterinary, chaplain's departments, post-office, military police . .	466	
Total all ranks	10,155	

With 2,450 horses, 18 guns, and 320 carriages.

There is in 1st line a transport company, which carries, among other things, one day's provisions for the men; and in 2nd line, half a transport company, carrying, among other things, a second day's provision for the men, and one day's forage for the horses.

Two or more divisions form an army corps. The established organisation in our service is:— *An army corps.*

3 divisions = 30,462 all ranks.

Cavalry brigade (3 regiments with 1 battery horse artillery) employed for special purposes, 2,342.

Corps artillery (3 batteries horse artillery, 2 batteries field artillery, and an ammunition column).

Corps engineers (1 company and field park, 1 pontoon troop, and half a telegraph troop).

Commissariat, ordnance store department, medical, &c., &c., comprising 8 transport companies.

Such a corps contains 21 battalions, 15 batteries, and 6 regiments of cavalry, numbering of all ranks :—

37,045 men, with 12,939 horses, 90 guns, and 1,573 carriages.

Ground in relation to Tactics.

Ground must suit the different arms.
Cavalry.
Artillery.

* Ground should be suitable to the arms of the service to be employed. Cavalry, if possible, should be screened by some elevation, with firm and nearly level ground for an advance.

Artillery ought to have high positions with cover, and soft or broken ground in the immediate front, and a clear range with hard surface.

Infantry.

* Infantry can cross any country. Undulations, hedges, farmsteads, &c., protect them in an advance. Non-commissioned officers and men must understand how to make use of these, and many other accidents of the surface; as by such means they can approach the enemy with precaution, find out his whereabouts, and report his movements. What a soldier has to learn is how to post himself so as to watch his enemy, while he himself remains concealed.

Importance of cover.

* Cover is of great importance in getting troops unseen into position before an action; or, during a fight, in transferring them from one part of the field to another; or to take the enemy by surprise. Hedges, which are not shot-proof, should not be greatly made use of, for, if men collect behind them, they may suffer severely from artillery. Yet they afford good concealment; and often it is possible to deceive an enemy as to the number of men behind a

NATURAL COVER.

bank or wall, if one or two men show the tops of their heads above it at different points.

The extent of front on which troops can move depends upon the ground; and the surface has much influence in regulating the pace. Such obstacles as streams, marshes, canals, &c., greatly delay an advance. But as time is so important in tactical operations, every effort should be made to anticipate these difficulties by extemporising means of crossing, and repairing bad places with materials at hand. Dikes and fences, though unimportant obstacles, give protection to infantry. The value of such shelter in the attack is surprising (see Plate I.). Thus men in extended order can take advantage of manure heaps, or haycocks in a field; and a mass of troops may be hidden behind a comparatively insignificant slope. Hence for the attack cultivated and accidented ground is best. But for the defence the country in front cannot be too open. Houses and villages can generally be defended.

* Broken ground best for attack. Open in front best for defence.

CHAPTER II.

ENCAMPMENTS, FIELD COOKING, SHELTER TRENCHES.

Encampments.

Site. * THE position of a camp depends first, on military, secondly upon sanitary considerations; the character, movements, and proximity of the enemy deciding which reasons weigh most in selecting the site.

Open ground is preferable. We should choose a sandy soil rather than clay, and avoid wet or marshy ground. The mounted services should have the most level ground. Cavalry and artillery are placed either in second line behind the infantry, or with infantry on their flanks for protection. The commissariat depôt should be established near a road.

Formation of camps. * Camps are formed of huts or tents—bivouacs of brushwood, straw, branches, or anything else handy. It is best to encamp in rear of a position to be occu-

Four requisites. * pied. Troops require these four necessities in the following order of importance:—(1), water; (2), wood; (3), forage; (4), straw.

Arrangements for water. * If water is furnished from a stream, the stream should be divided into three portions; the upper for the men to drink; the middle for horses and cattle; the lower for washing. Sentries from the nearest

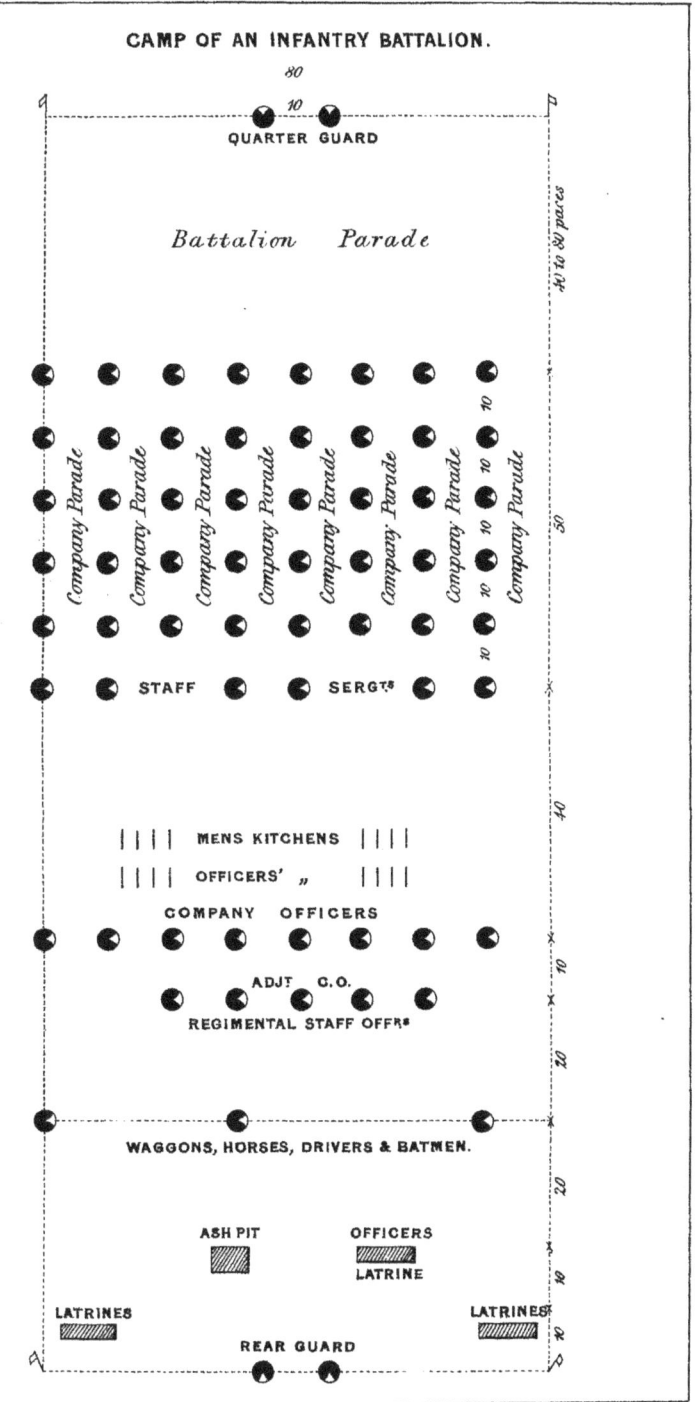

Plate 11.

guard, or a special water-guard, must have strict orders to enforce this arrangement. In peace time, wood, like rations, is issued on the ground.

The usual form of camp for an infantry battalion is that shown in Plate II., *Regulations for Encampments*, of which Plate II. is a copy. 10 paces is a convenient and sufficient interval between tents and companies. Each bell tent holds 15 men, but 12 is the usual number. The colour-sergeant occupies the rear tent, to be near his captain. * Camp of a battalion.

If the camp is to stand some time, tents should be struck and repitched every 2 or 3 days, arms and blankets removed, and the ground well brushed with boughs. Tent-flys should be rolled up the first thing in the morning, but in wet weather only on the leeward side. As a rule the doors face the head of the column. It may be ordered otherwise, but all must face alike. Drains are cut the width of a shovel, and a few inches deep, just outside the curtains; from which others are cut, leading into main drains between every second row of tents, or in a different direction, according to the natural slope of the surface, which is best seen by watching the flow of water after the first shower of rain.

The quarter-master and 4 camp colourmen march with the general advanced guard, or some distance in front of the column. One of the party should know the strength of the battalion. A staff-officer will mark out the ground, and dress two camp colours on the front line of the camp. 80 paces is sufficient front, and 200 paces for depth, allowing 40 to 80 paces for parade ground. The quarter-master then proceeds to lay out the camp, placing the rear camp * Laying out the camp.

colours at right angles to the front, by means of a cross staff, aligning one pair of sights with the front line, and then looking through the other pair to fix the camp colour; or a right angle may be laid out with a tape held at 3 feet, 4 feet, and 5 feet, or any multiples of these, and stretched on the ground (*vide* Fig. 1), because $5^2 = 3^2 + 4^2$; or by means of a cord with the two ends held along the line, while another man holding the centre stretches it on one side, and then passes it over to the other (*vide* Fig. 2).

The position of the company and officers' tents are quickly fixed by means of small iron pegs with yellow vanes on them, attached to a cord at intervals of 10 paces. The front of each row is marked with a small yellow flag, with the letter of each company. The quarter-guard is fixed in the centre of the front line of the parade, its doors turned to the front. The rear guard is usually in the centre of the rear line of the camp turned towards the rear.

Arrival of battalion. * A mounted officer, having ridden forward to find the camp, returns to conduct the battalion to the ground. On arrival on the parade ground, it is formed in column, with each company in prolongation of the line marked out for its tents. The band, drummers, pioneers, and pipers join their companies; arms are piled, and the accoutrements taken off and hung on the piles of arms. The colour-sergeant tells off the company into squads, according to the number of tents, with a N. C. officer in charge of each. From every squad, 1 N. C. officer and 6 men (1 file polemen, 1 file pegmen, 1 file packers) are told off ready to pitch tents when the waggons arrive. The adjutant details the following parties:—

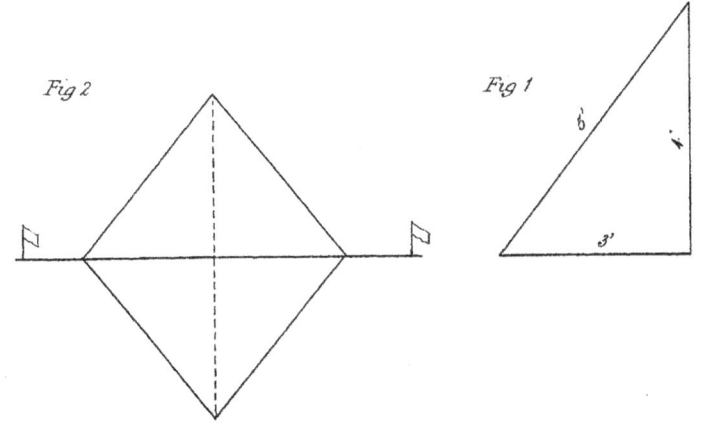

BIVOUAC OF BRANCHES

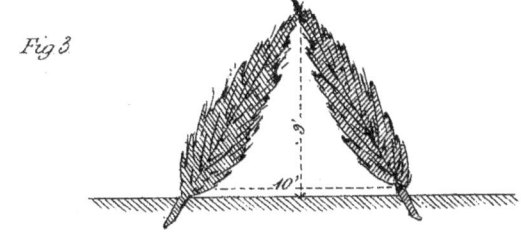

SECTION OF BIVOUAC

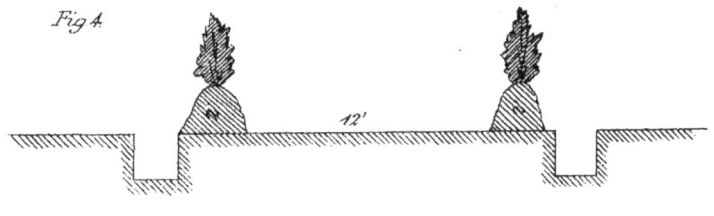

i. Cooking party—2 men per company under the sergeant cook, who lose no time in making the kitchens. *Parties told off.*

ii. Latrine party—All pioneers with picks and shovels, and 2 men per company. These dig latrines in sheltered places when possible.

iii. Water party—1 N. C. officer and 2 men per company, under a sergeant.

iv. Ration party—1 N. C. officer, and 2 or more men per company, under the quarter-master-sergeant, who go to the depôt for rations.

v. Wood party — 2 or more men and a N. C. officer per company. These collect wood and bring it to the kitchen, if no fuel is carried.

The rest of the men lie down by the arms.

A field officer mounts the quarter and rear guards.

The baggage-waggons are conducted to the rear of the camp by an officer, who takes care not to pass over the camping-ground, but through the battalion intervals. They are halted with the tail-boards to the front along the rear line. A subaltern marches 5 men out of each tent party to the waggons, unpacks, and brings up the tents. *Arrival of baggage.*

The other men of the tent parties, viz. front rank polemen, are paraded by the captains in rank entire on the reverse flank of the companies, the nearest men to the companies covering between the camp colours. The captains extend the polemen to 10 paces, and dress them from front to rear. At the same time a mounted officer dresses them from flank to flank. *Extension of polemen.*

These men keep the exact spots for the tent poles.

Tent pitching. * The parties bring up the tents, pegs, and poles. A peg is driven in at an angle at the heels of each poleman to assist in raising the pole, the two pieces of which are put together and the top shipped into the cap of the tent, which is stretched out flat with the door hooked up. 4 men hold the storm guys—if there are no guys they hold the 4 red runners. When all is ready the bugle sounds one G, and all tents are raised together. The guy ropes are first fixed, and the tent is pegged down, working round from right to left, each rope being pulled out in prolongation of the seams of the canvas. The arms, accoutrements, and blankets are then brought in and drains cut.

The tent doors face the front as a rule, unless they are turned in another direction on account of weather.

The arms should be stacked round the pole. Sometimes rough arm-racks are made. Articles may be hung from the pole by cords attached with a clove-hitch.

For instruction it is well to form up the parties and change rounds, so that No. 1 becomes No. 2, No. 2 becomes No. 3, and No. 3 becomes No. 1, for striking and pitching tents again.

Striking tents. * In striking camp the blankets are first rolled and packed. After breakfast the trenches, ashpits, and latrines are filled in, kitchens and their chimneys are levelled, and all fires put out. Tent-pegs are carefully pulled out, and all spare ones placed in the bag. Doors are hooked up, and everything made ready for lowering when the G sounds. Then all the tents are let down together, doors uppermost, each

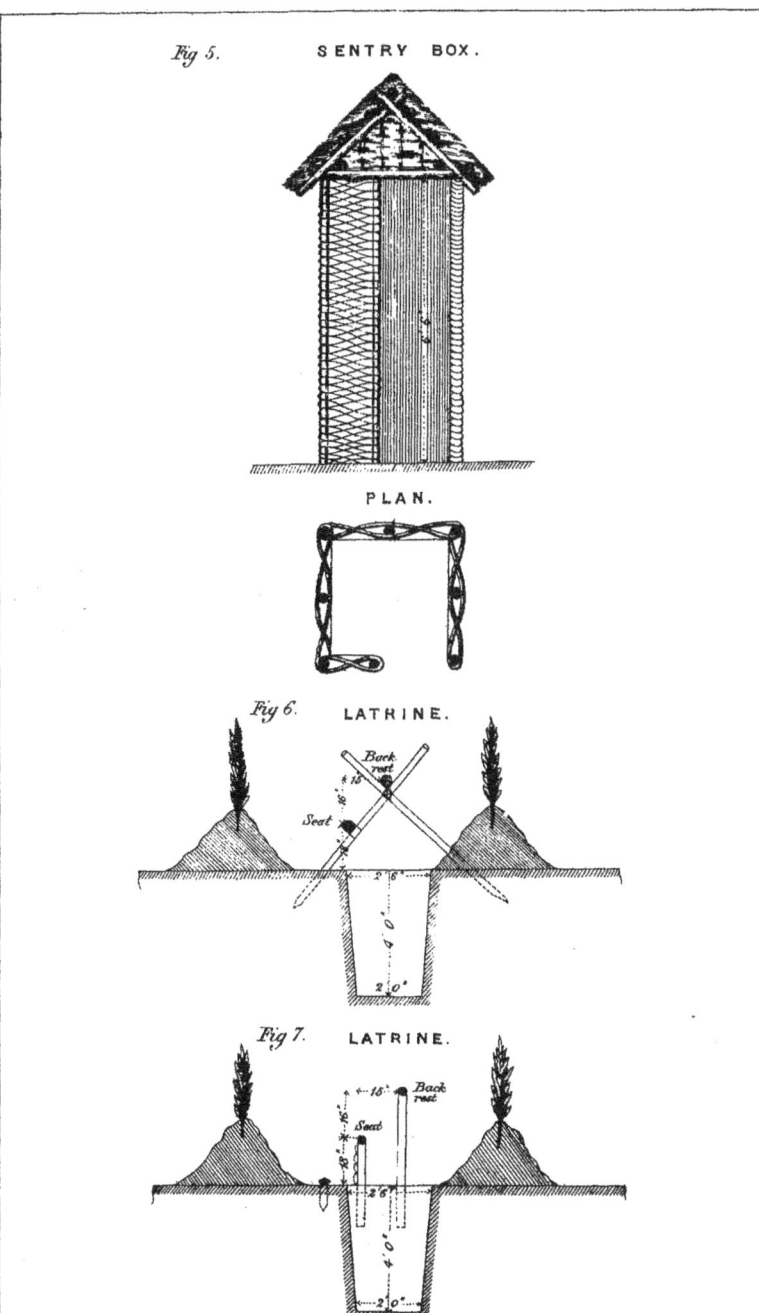

tent stretched out smooth and flat, folded neatly and small three or four times, so as to be placed easily in its bag. The mallets and remaining pegs are collected and all stowed in the waggons.

A bivouac is formed with the same regularity as a camp. The arms are piled in quarter column, and men lie down in line with them to one flank. *Bivouac.

Hurdles, ferns, and dry leaves make a rough bed. Something should be got to raise the head off the ground; and the hint to excavate a hollow for the hip is not to be despised.

A row of bushes or branches helps to keep off the wind. Figures 3 and 4 are examples of such shelters. If earth is required to form a mound it should be taken from the outside, and the interior ground should not be disturbed.

A sentry-box can be improvised with a few poles, such as hop-poles, with straw, wicker-work, or bushes interlaced (*vide* Fig. 5), or by means of two pairs of waggon shafts leaning against each other, and covered with a tarpaulin. Failing such expedients, sentries may use any natural shelter near their posts. *Sentry box.

Latrines are dug by the party told off under the pioneer-sergeant without delay. Unless orders are given to keep them within the limits of the camp, they may be made in some convenient sheltered place. The simplest form is a trench 15 feet long and 1 foot 6 inches deep, sufficient for one night. But if the camp is for a longer time, it must be deeper; and, if spars are available, two cross poles are firmly driven in at each end to support a back rest, and a seat lashed to them (*vide* Figs. 6 and 7, *Latrines.

copied from the *Regulations for Encampments*). The earth is heaped up on each side, covered with bushes or small trees for shelter. Sometimes strips of canvas are provided for shelter. Latrines must never be dug near the water supply. A few inches of earth should be thrown over the soil daily.

Loading baggage waggons. * In order to load the baggage-waggons, one man gets up into each waggon, takes off the cover, and packs. The rule to remember is that the articles required last are to be put in first. Thus, the blankets are placed at the bottom, then the tents, and, lastly, the tools and wood.

The driver stands to his horses' heads; on no account is a waggon to move until the cover is replaced and properly secured.

The waggons should march in the order companies stand on parade. The letters of the companies should be chalked on the waggons, or else a small flag with the letter carried on the front of each.

Baggage guard. * The baggage guard is usually formed of the officers' servants, under a N. C. officer. Their duty is to maintain order among the drivers, and to see that nothing is lost or taken from the waggons. They march with fixed bayonets—a corporal and two men in front, the rest distributed on both sides of the convoy, the N. C. officer in charge following in rear. On no account are they to ride, or place their arms or valises on the waggons.

Cavalry camp. It is well to know that cavalry prefer the method of encamping in column of troops at half-order, as shown in Plate VI., *Regulations for Encampments.*

Artillery camp. And the plan recommended for a temporary artillery camp is that shown in Plate X., *Regulations*

TORRENS KETTLE   FLANDERS KETTLE

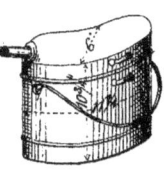
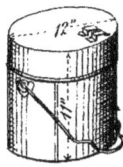

SUNKEN FLYING COLUMN TRENCH

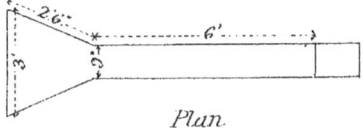

WALL TRENCH

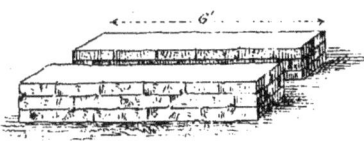
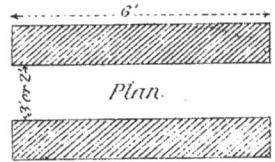

RAISED FLYING COLUMN TRENCH

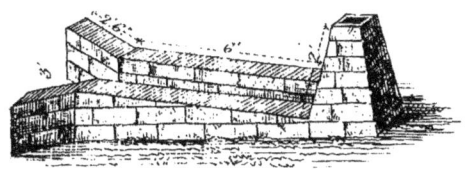
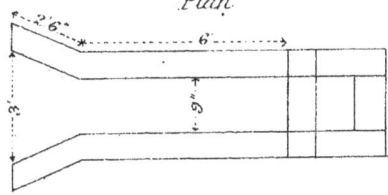

for Encampments, in column of subdivisions, using the carriages instead of picquet posts, and fastening the picquet lines to the wheels. This is the quickest and safest way of securing the horses. The carriages are always kept limbered up, and the harness buckled together in rear of the pair of horses to which they belong.

Field Cooking.

There are two kinds of kettles in use (1), the "Torrens," which is bean-shaped (*vide* Fig. 8). It weighs 3 lbs. and cooks for 5 men, or without vegetables for 8 men. Torrens kettle.

(2) The "Flanders," which is cylindrical (*vide* Fig. 9). It weighs 8¾ lbs., and cooks for 8 men, or without vegetables for 15. Flanders kettle.

The following is a brief description of the trenches generally employed :—

i. The sunken flying-column trench (*vide* Fig. 10), used when a battalion encamps for a day or two. The chimney, 2 feet high, is formed of sods cut off the top of the trench. Time to make, 2 men in 15 minutes. Time to cook, 1 hour. It should be made, if possible, on a gentle slope, the mouth facing the wind. It holds 6 Flanders or 9 Torrens kettles, and cooks for about 50 men. Therefore, as a rule, one of these trenches is required per company. Sunken flying-column trench.

ii. The wall trench, suitable for wet or marshy ground (*vide* Fig. 11). For the Torrens kettle it is 18 inches high, and 2 feet between the walls. For the Flanders kettle it is 2 feet high, and 3 feet Wall trench.

between the walls. It is built with sods, the largest placed at the bottom. Time to erect, two men 30 minutes. Time to cook, 1 hour. The kettles are slung across on bars of wood or iron. It holds 12 Flanders or 18 Torrens kettles. It will cook for 2 companies of 50 men each; therefore, as a rule, 4 are required for a battalion.

Raised flying column trench.

iii. The raised flying-column trench (*vide* Fig. 12), also adapted for wet or marshy ground. It is built with sods, cut close by. Time to erect, 2 men 30 minutes. Time to cook, 1 hour. It will hold 6 Flanders or 9 Torrens kettles, and cooks for a company of 50 men.

Broad arrow kitchen.

iv. The broad-arrow kitchen (*vide* Fig. 13), adapted for standing camps. The base of the chimney is 3 feet square; its height 5 feet; and at bottom it is 1 foot square. Each trench holds 9 Flanders or 11 Torrens kettles. The kitchen cooks for 165 men with the Torrens and 220 with the Flanders kettles. Time to construct, 5 men 4 hours. Time to cook, 1 hour. Each trench is 18 inches deep at the mouth, and for 18 inches inwards, then it slopes gradually up to 6 inches where it enters the chimney. All trenches should be cut, if possible, on a gentle slope, with their mouths towards the wind. A stiff plaster is made of clay, and a kettle used as a mould, round which the clay is placed to cover the trenches. When dried the cracks are filled in. One or two of the handiest men should do this work.

Triple arrow kitchen.

v. The triple-arrow kitchen is adapted for standing camps—*e.g.* a militia regiment out for training (*vide* Fig. 14). It will cook for about 600 men. Time to make, 12 men 10 hours. Time to cook,

BROAD ARROW
KITCHEN

Fig. 13.

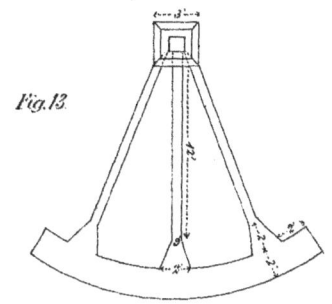

SECTION OF ABOVE

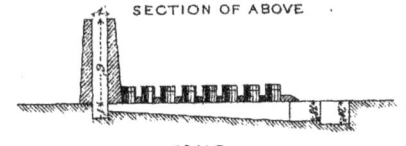

SCALE

TRIPLE ARROW
KITCHEN

Fig. 14.

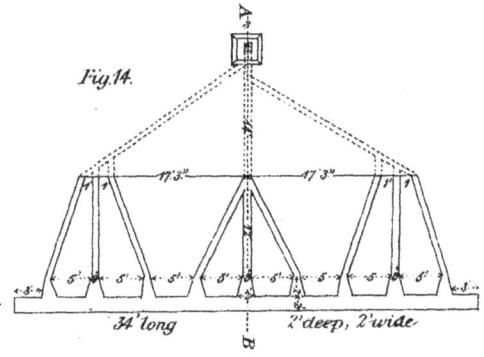

SECTION ON A.B.

SCALE

1 hour. The trenches are 18 inches deep at the mouth, sloping upwards to 9 inches at the top, covered with clay, with places moulded for the kettles. The base of the chimney is 3 feet square; its height 6 feet; and the bottom 1 foot square and 1 foot deep.

The marking out of all these kitchens is much facilitated by having a bundle of pegs, with twine attached at the different dimensions.

vi. Lastly, the open-hole trench (*vide* Fig. 15), used for men to cook separately in their canteens, is made with sods or stones, or a hole may be dug in the ground, according to circumstances. *Open hole trench.*

Iron tripods, 2 feet high, supporting a rod from which the kettles are suspended, have been used in manœuvres by some regiments. They are very portable, and the dinners require a shorter time to cook, but more fuel is wanted than when trenches are dug. *Tripods*

These short notes should be borne in mind:—

(1) Three pounds of fuel is allowed per man per diem—1 cubic foot of wood makes 10 rations. *Rules for cooking.*

(2) Water should boil in 20 minutes after the fire is kindled, and dinners should cook in an hour, which is sufficient time under difficulties. If wood is at hand and dry, they will be ready sooner.

(3) Meat should be cut into pieces of ½ lb., placed in the kettles, and simmered slowly until cooked. Vegetables, such as onions, may be added.

(4) Potatoes may either be cooked separately or with the meat; but if not pealed they must be cooked separately.

Shelter Trenches.

Description of trench.
Shelter trenches are shallow excavations of ground, sufficient, with a parapet, to cover troops in line kneeling or lying down. The front rank acts as a covering party, and detains the enemy, if possible, until the trench is completed, when they, with the rear rank, man the trench, and open fire.

Necessity for intrenchments.
They are necessary, both in the attack and in the defence, owing to the great range of infantry and artillery fire, in situations where natural cover cannot be utilised (*vide* Plate III). For this reason, troops must have intrenching tools in the field, whether carried as part of their equipment, or conveyed in carts. The drill is laid down in the

Working parties and laying out tools.
Field Exercises. Working parties are detailed of complete companies, battalions, or brigades. Tools are laid out, as in Fig. 16, at 1 pace interval, the rows being in quarter column at 6 paces distance. The companies for work are formed in quarter column to one flank.

Drill.
After opening the ranks and causing the slings to be loosed and arms advanced, the captain orders—"For shelter trench exercise, sling arms," which is done in two motions, as in *Rifle Exercises,* Page 48. The rear rank is turned to the left (or right), and the order given to "file on tools," upon which the marker runs to the far flank of the tools.

After halting and fronting his rank, the left guide orders the men to "take up tools," when each man takes a short pace with the left foot, and, stooping down, takes a pick in the left hand and a shovel in

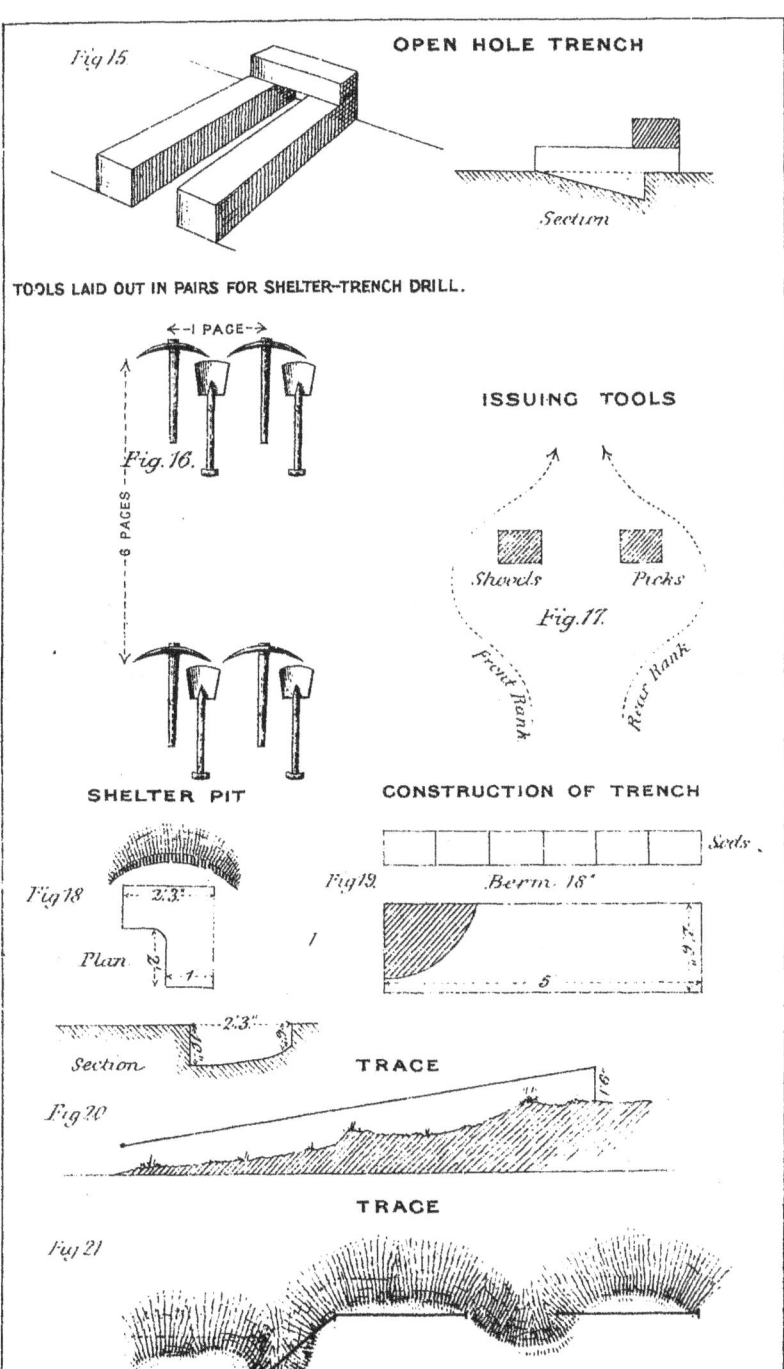

Plate III

ARTIFICIAL COVER.

the right, coming to the trail without noise, and carrying the iron to the front and vertical.

The guide turns the men to the right (or left), and in turning they must drop the head of the pick and raise the head of the shovel. When in file they incline the handles outwards, to enable them to close up. He marches the rear rank back, and they are halted, fronted, and dressed.

In order that each rank may carry half the tools, when the captain orders "Transfer tools," the rear rank give the front rank the shovels; the latter take them with the right hand, and pass them in front of the body into the left hand; then both ranks carry the tools in the left hand. *(Marching to the ground.)*

If the tools are in heaps, or in a cart, two non-.com. officers issue them; shovels to the tfron rank and picks to the rear rank, both ranks filing past the tools (*vide* Fig. 17), and closing in again, the leading men stepping short. *(Issuing tools from heaps or a cart.)*

Approaching the ground, the front rank hand the rear rank the shovels at the command "Transfer tools." The right guide extends the front rank as a covering party. They lie down 150 paces in front; and if there is no natural cover available, and time permit, they make shelter pits, *i.e.* shallow excavations for the use of skirmishers for a short time; each pit taking 5 minutes (*vide* Fig. 18). For this 1 pick and 1 shovel are allowed to each group of 4 men. They can be improved by making loopholes of sods. If necessary, and time permit, they may be converted into rifle pits, which see (Page 184). *(Extension of covering party.)*

The rear rank open to 2 paces interval, and halt

<small>Extension of working party.</small> 12 paces in rear of the line selected for the trench. They ground arms, butts to the front, so as to be handy in a moment, and take off jackets and purses. An officer steps along the line, giving 2 paces in length to each man, who fixes his pick in the ground on the left of his task, the shovel along the line, blade to the left, and then lies down till the extension is completed.

<small>Construction of the trench.</small> Word is passed when all is ready, and all set to work to cut and build up the surface sods along the line, leaving a berm of 18 inches. Each man excavates a hole of the required depth at the left corner of the task, turning to the left in working (*vide* Fig. 19). If the ground be hard, grooves must be cut to the rear, and the ground undermined with the pick.

<small>Guide to measurements.</small> As a guide to measurements—the handle of the pick-axe is a little over, and the shovel is a little under 2 feet 6 inches. The top stud of the shovel is 1 foot 6 inches. But new pattern tools must be proved. A sword blade is 3 feet.

The length of trench which can be made is twice the number of files in paces. Thus 100 files will make 200 paces. Intervals should be left for the passage of cavalry and artillery. Roads ought not to be cut up. As a rule, the shelter trench follows the crest line of the hills, sufficiently near the brow to sweep all the ground in front with fire. The <small>Trace and fitting the trench to the ground.</small> chief point to determine is how high up the slope, or how low down, the crest of the parapet must be. If too high, the slope is not "seen"; and an enemy can assemble in comparative safety to rush the trench. If too low, the trench is seen into from opposite heights, there is no protection, and if the defenders have to retire, they must run some yards up the hill exposed to full view.

The practical way to fit the trench to the ground is to place sword blades or cleaning rods at intervals along the trace, while another person descends the slope some distance, and, lying down, makes sure of seeing the tops of the swords or rods. Fulfilling these conditions, the shorter the trench the better, to save labour; and by making it as much as possible parallel to the general front, there is less danger of suffering from enfilade fire (*vide* Figs. 20 and 21).

The first task in half an hour is 2 paces (5 feet) × 2 feet 6 inches × 1 foot 6 inches, and shelters one rank kneeling in the trench and one rank lying down behind.

Tasks.

The second task, in half an hour, 5 feet × 2 feet 6 inches × 1 foot 6 inches, shelters both ranks kneeling in the trench.

The third task, in 1 hour, 5 feet × 3 feet × 1 foot 6 inches, shelters both ranks and the supernumeraries lying down in the trench.

Thus the trench completed requires a task from each man 5 feet long × 8 feet broad × 1 foot 6 inches deep. The berm which is 1 foot 6 inches, is unavoidable. The parapet is 1 foot 6 inches high. At 200 yards a bullet penetrates 12 inches; therefore, do the utmost to raise a bullet-proof parapet quickly, cutting the surface sods thick and square, and building them along the berm; then placing every shovelful of earth just in front of them, and not scattering it. As the work proceeds, the earth is thrown more forward, taking care that the parapet is never more than

1 foot 6 inches high, with the inner slope as steep as possible, and the upper slope like a glacis, shovelled down to coincide with the natural fall of the ground. The men must throw the earth towards a salient angle, where there will be a deficiency; but at a re-entering angle they must throw it outwards, for there will always be a surplus.

Heather and bushes strewed over the parapet greatly help to conceal the newly-made work (*vide* Fig. 22).

Best diggers in rear rank.

The strongest and best diggers are placed in the rear rank. They perform the first and third tasks, the front rank the second. The whole work, including the sending out and withdrawal of the covering party, must be conducted in strictest silence. At the completion of each task the men lay their picks and shovels together behind their work, handles to the front, and then dress.

Withdrawal of covering party.

The men of a covering party, when called in, must retire straight back, jumping the parapet without injuring it.

The exercise should be concluded by manning the trench and firing a few volleys. The instant a volley is fired the men must lower their heads behind the parapet.

The charge.

Bayonets are fixed in the trench, and all charge over the parapet.

Tools must be returned to the cart, or deposited in rows without noise, and according to the drill.

Charger pit.

Sometimes shelter is required for officers' chargers, as when no natural cover is near. A charger pit is made in a convenient spot, in rear of the shelter trench, concealed as much as possible. It takes 3

TASKS

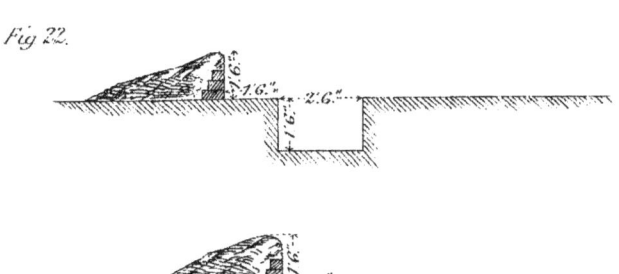

Fig. 22.

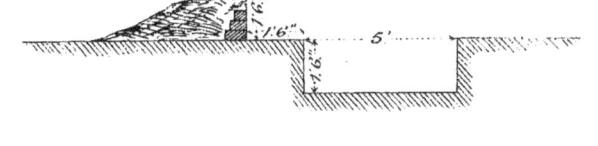

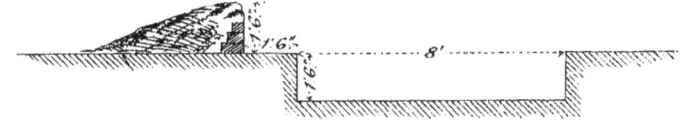

CHARGER PIT

Fig. 23.

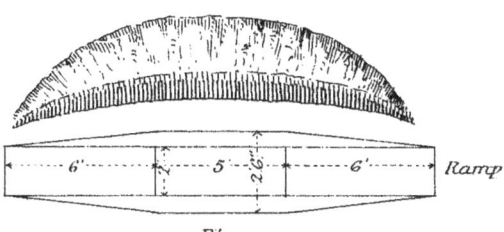

Plan.

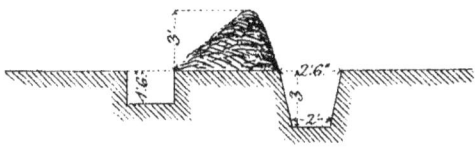

Section.

men 2 hours—1 man digs in the trench, 1 gets earth from a ditch outside, 1 man builds up sods, and picks when required. A N. C. officer superintends (*vide* Fig. 23).

The duty of filling in trenches should be performed by a fatigue party, generally by defaulters. The N. C. officer in charge of them must see the work thoroughly done, and the earth well trodden and beaten down, otherwise the ground will be dangerous for horses.

Filling in trenches

CHAPTER III.

THE THREE ARMS.

Cavalry. Not only is it necessary to learn the characteristics
Necessity and functions of our own arm of the service, but
for some
knowledge we must have some knowledge of the others, in
of the
other order to appreciate their assistance, to understand
arms. how to co-operate with them, and the best way to
encounter them as foes.

Duties of The duties of cavalry are to screen and cover
cavalry. the advance of an army, as also a force halted, or in
retreat; to reconnoitre, gaining information of the
Detached. country and of the enemy. If cavalry watches and
feels for the enemy thoroughly, is prompt in securing
information of his movements, and hangs constantly
on to his troops, a most important service will be
rendered to the army, and proportionately demoralising to the enemy.

With this object independent parties move far ahead of the main body.

Cavalry also furnish escorts and orderlies.

The above are termed detached duties.

On the On the field of battle cavalry must seize every
field of
battle. opportunity for an effective attack, following up and
confirming success, and, if needs be, it must sacrifice

itself to avert disaster. Its leading principle is the attack, and even in defence it must advance to attack, being comparatively defenceless halted. Hence it should be kept out of fire till the proper moment. To be effective the advance must be rapid and unexpected, the charge well timed and impetuous. Its great tactical value depends on mobility and celerity combined. After a charge cavalry pursues or rallies rapidly.

In covering a retreat the only way cavalry can defend the other arms is by continuous and vigorous offensive action, often suffering disadvantages as to time and place, and even at the risk of complete destruction.

Covering retreat.

A cavalry soldier mounted in attack or defence depends upon his sword or lance, not on his carbine, which should be used from the saddle for signalling only.

Arms.

The effective action of cavalry is of three natures— (1) Shock action in line, *i.e.* the charge; (2) detached action in small parties or singly; (3) dismounted fire action. The latter can be used advantageously under certain circumstances, chiefly on the defensive, *e.g.* to check an advance, or to hold a post, or sometimes a party of cavalry is pushed well to the front to seize and hold some point until the infantry come up.

Three modes of action.

During the recent war between the Russians and Turks, and later in Egypt, mounted infantry were useful when there was not sufficient cavalry. Very similar was the cavalry raised during the American Civil War, when it executed remarkable raids. But there is this danger in employing a mounted force so improvised against an army well supplied with cavalry,

Mounted infantry.

that it will be neither one thing nor the other—indifferent infantry and indifferent cavalry.

Tactical unit. The tactical unit is the squadron. The largest tactical body is the division, composed usually of two brigades with horse artillery attached.

A cavalry brigade consists of 3 regiments and 1 battery of horse artillery attached. Each regiment has 4 squadrons. The right and left half of each regiment are termed respectively right and left wings. Two troops form a squadron. A regiment of cavalry **Formations in the field.** moves in the following formations:—In line; open, quarter, and close column of squadrons; open column of troops; column of fours, sections, half sections, files and single files.

Speed. The speed is at a walk 4 miles, at a trot 8 miles, at a gallop 12 miles an hour.

Column of fours is eight abreast, 4 front rank and 4 rear rank.

Column of sections is 4 front rank, covered by 4 rear rank.

Column of half-sections is 2 front rank covered by 2 rear rank.

Frontage. In calculating the frontage of cavalry in line, allow one yard for every trooper in the front rank + 12 yards for each squadron interval + 12 yards for each regimental or brigade interval, or more for the **Regiment in line.** band and staff if required. Thus a regiment in line of 4 squadrons, each of 96 sabres, will require 228 yards; for each squadron has 48 horses in the front rank, and $48 \times 4 + 12 \times 3 = 228$ yards.

Brigade in line. Or, suppose a brigade, of 3 regiments, each of 400 sabres. Every squadron then has 50 in the front rank; the front of each regiment will be $50 \times 4 +$

Fig 24.

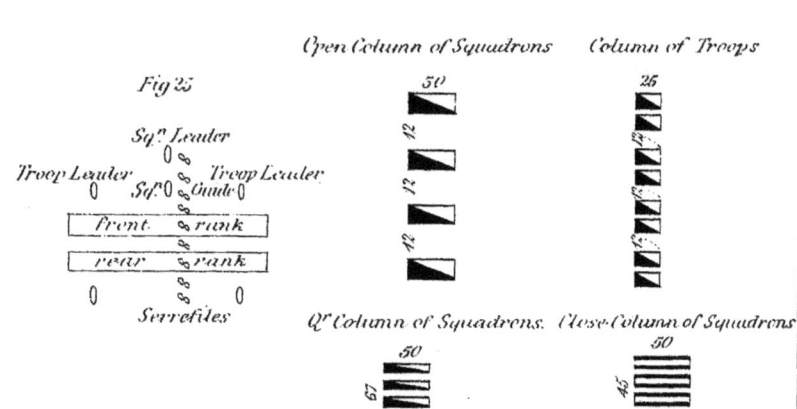

Fig 26.

Fig 25.

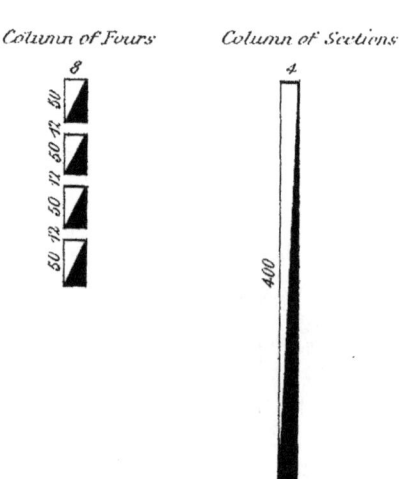

Fig 27.

$12 \times 3 = 236$; and the front of the brigade $236 \times 3 + 24 \times 2 = 756$ yards, (*vide* Fig. 24).

As a horse's length is 8 feet, the depth of a squadron in line is readily seen to be $8 \times 9 = 72$ feet $= 24$ yards by reference to Fig. 25. Squadron.

The depth of a regiment in open column of squadrons, say each of 50 sabres $= 50 \times 4 + 12 \times 3 - 50 = 186$ yards. Depth in open column of squadrons.

The depth of a regiment in column of troops $= 50 \times 4 + 12 \times 3 - 25 = 211$ yards (*vide* Fig. 26). Column of troops.

To find the depth of a regiment in quarter column of squadrons, allow 6 horses' lengths for each squadron, and add 1 horse's length. The depth will then be 25 horses' lengths $= \dfrac{25 \times 8}{3} = 67$ yards. Quarter column of squadrons.

To find the depth of a regiment in close column of squadrons, allow 1 horse's length between the rear rank and the front rank of successive squadrons, and add 1 horse's length. The depth will then be 17 horses' lengths $= \dfrac{17 \times 8}{3} = 45$ yards (*vide* Fig. 26). Close column of squadrons.

In a column of fours, the squadron distances are preserved (*vide* Fig. 27), the depth = the front in line + squadron intervals $= 200 + 36 = 236$ yards. Fours.

If the same regiment were in column of sections, the depth of the column in yards = number of horses in the ranks $= 400$ (*vide* Fig. 27). Sections.

Squadron distances are not maintained with a less front than fours, for cavalry never manœuvre with a less front than this; and on emerging on a plain from column of route, they will assume fours, or other formation, without delaying the rear.

It is obvious that the length of the regiment in

Half sections.	column of half sections in yards = double the number of horses in the ranks = 400 × 2 = 800 yards; and in single files it will be four times the number of horses in the ranks = 1600 yards.
Interval and distance between cavalry and infantry.	The interval, as well as the distance, between cavalry and infantry is 25 yards.
War establishment of a squadron.	The war establishment of a squadron is 6 officers, 2 troop sergeant-majors, 6 sergeants, 8 corporals, 4 artificers, 2 trumpeters, and 120 privates, with 2 drivers—total 150 officers and men; 18 chargers, 120 troop horses, 4 draught horses—total 142 horses. If we deduct horses to mount sergeant-majors, sergeants, artificers, and trumpeters, and allow for a few casualties, the squadron will consist of 48 to 50 files or 96 to 100 troopers.
Tactical division of squadron, and how numbered.	The squadron is divided tactically into 2 troops. The senior of the two captains on parade is the squadron leader.

Squadrons are numbered from the front in column, and from the right in line. The two troops of each squadron are termed "right" and "left" in line, and "leading" and "rear" in column.

War establishment of regiment. The war establishment of a cavalry regiment is 31 officers, 600 non.-com. officers and men, 22 drivers for regimental transport—total all ranks, 653; 91 chargers, 480 troop horses, 44 draught horses—total 615 horses.

Cavalry manœuvres in column with a front depending on the ground. Small columns are flexible and mobile, offer a small mark to artillery, are more easily sheltered, and able to turn obstacles.

Formation for attack. Its formation for attack is in echelon, in successive lines, with a reserve in rear of one flank, or both, to protect retreat in case of failure, or to follow up victory if successful. The greatest care should be taken to protect the flanks; and before advancing to the charge, scouts should reconnoitre the ground to the front and flanks, to see that it is not marshy, or intersected by dikes or other obstacles. *Flanks vulnerable.* *Scouts.*

Cavalry ought to attack cavalry on flank, because the flanks are its weak points; or when in the act of deploying. Favoured by such a chance of surprise, a small body may attack a larger with every prospect of success. *Cavalry against cavalry.*

It can easily attack artillery in motion, or when limbering or unlimbering. Guns in position should be attacked in flank and in rear, a portion charging the escort at the same moment. Sometimes a converging attack may be made by skirmishers in front. *Against artillery.*

Cavalry will not usually attack steady infantry, except by surprise. The best chance is when the infantry is moving; or has an exposed flank; or is demoralised or broken by artillery fire. *Against infantry.*

When acting in support of infantry, cavalry should be so disposed as to prevent the most advanced troops being scattered by the enemy's cavalry. *Supporting infantry.*

The general duties of artillery are to commence and carry on an action at long range. During an attack it covers the deployment of the advanced guard, and assists in driving in the enemy's advanced troops. If provided with a sufficient escort, it may surprise an enemy, and shell an unguarded camp. On the defensive, it checks the enemy's deployment, obliging him to form up at a distance, and thus delays him. It is most useful in a recon- *Artillery. Duties of artillery.*

naissance in force, causing the enemy to disclose his strength.

During an engagement, it keeps down the fire of the enemy's artillery and infantry, searching his position, and bringing the guns to bear successively on all approaches, woods, ravines, and all made or natural cover.

In the attack. Supporting an attacking force, guns will play upon the point of attack, and change position as their fire becomes masked, guarding against the enemy's skirmishers.

In the defence. In a defensive position, they keep down the enemy's artillery fire, delay the advance of his infantry, protect the flanks, and assist in any counter-attack.

Artillery, also, co-operates with cavalry and infantry in striking a final blow, following up a retreating fire if victorious. In case of retreat it covers the retirement.

Principle of employment. The principle is to employ artillery against that arm of the enemy which, for the time being, is most prominent. During the different stages of a battle, one arm acts as the principal, therefore every effort must be directed to destroy it.

Artillery fire should rather be directed against troops than guns. Great advantage is gained by concentrating fire on the enemy's artillery at the beginning of an engagement, and so clearing the way for the infantry; and before an assault, artillery should direct its fire so as to disorganise and distract the enemy.

In a position attacked, guns should be directed against the supports, leaving the fighting line to the infantry; for the supports offer a better target, especially while in column; and it is most important to check and weaken them, for then they will be less ready to push on; while the fighting line, if unsupported, will lose heart, and go no further; or should they press on, their weak attack will probably fail. Advantage must be taken of commanding positions to destroy parapets, palisades, buildings, &c. *Fires at supporting troops when in defensive position.*

Artillery is that arm of the service which deteriorates least in action. It can be kept most effectively in hand by a commander, and can move rapidly.

It can be employed at ranges where infantry are useless. It has the advantage of curved fire possessed by no other arm; and is superior in physical and moral effect. Artillery reserves can be kept under cover till needed. Artillery is bulky and expensive, difficult to train and equip, and powerless in movement. It cannot be used alone, and is often insufficient for its own protection. *Power of artillery.* *But often helpless in motion.*

It has "fire action" only; and its weapons are:— *Weapons.*

Horse artillery, 9 and 13 pounders.
Field artillery, 9, 13, and 16 pounders.
} Rifle muzzle-loading.

Mountain guns, 7 pounders, 4 in each battery, carried on pack animals, are used in a difficult country, where there are no practicable roads. Field guns are mounted on sledges in winter in North America. Sometimes Hale's rockets are added to the equipment.

The tactical unit of artillery is the battery of

Tactical unit.	6 guns, which is subdivided for tactical purposes into :—
Subdivision of a battery.	i. A battery of 4 guns, and a division of 2 guns. ii. Two half-batteries of 3 guns each. iii. Three divisions of 2 guns each. iv. Six subdivisions of 1 gun each. But no smaller fraction than a division should be detached from the battery.
Speed.	The pace of artillery is, at a walk, 4 miles; at a trot, 8 miles; at a gallop, 12 miles an hour. Field artillery will travel on fair roads 4 miles; horse artillery 5 miles an hour.
Limit of opening for guns.	The narrowest space through which guns can pass is $6\frac{1}{2}$ feet.
Intervals and distances.	Intervals and distances are measured in action from muzzle to muzzle, and limbered up from No. 1 to No. 1. In action the distance between the trail-plate eye and gun leaders' heads = 10 yards. When the battery is in line at close interval there are 4 yards between subdivisions. When in line at full interval, with 6 horses, each gun interval is 19 yards, and the battery or brigade interval ($1\frac{1}{2}$ gun intervals) = $28\frac{1}{2}$ yards; with 8 horses, each gun interval is 23 yards, and each battery or brigade interval $34\frac{1}{2}$ yards.
Front of field battery in line.	Thus the front of a field battery in line with 6 horses = 5 subdivision intervals + 3 yards the front of one subdivision = 5 × 19 + 3 = 98 yards (*vide* Fig 28).
Front of horse artillery battery in line.	The front of a horse artillery battery in line = 5 subdivision intervals + 7 yards the front of the detachment = 5 × 19 + 7 = 102 yards. The distance from one gun or waggon to the next is 4 yards.

Fig. 28.

BATTERY IN LINE, FULL INTERVAL.

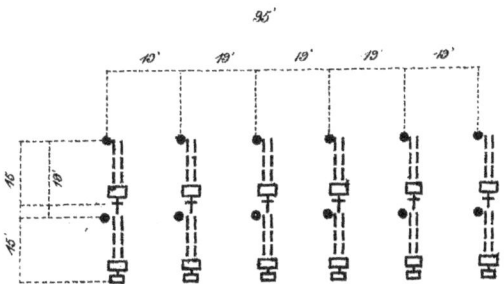

BATTERY IN ACTION WITHOUT WAGGONS.

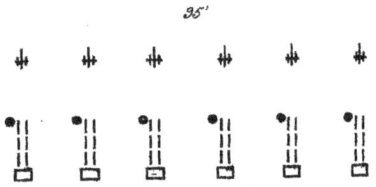

Intervals between artillery and either cavalry or infantry in line are each $1\frac{1}{2}$ subdivision intervals = $22\frac{1}{2}$, $28\frac{1}{2}$, $34\frac{1}{2}$ yards, according as the number of horses are 4, 6, or 8. *Intervals between artillery and cavalry and infantry.

In selecting an artillery position we must endeavour to obtain as many as possible of the following requisites:— A perfect artillery position.

The greatest available sweep of range both to the front and flanks. Guns should command not only the ground to the immediate front, but also the roads along which the enemy must advance to attack, and all exits from villages and defiles. But there is a limit to the range; for even with a good telescope the effect of shell cannot be seen beyond 3,000 yards. Wide range.

The height should not be excessive, for the guns cannot be depressed more than 10°; and guns on high ground are much exposed if placed forward on the crest to depress the muzzles. A gentle slope of 15° is best; for if steeper, the fire will be too plunging, especially at short ranges, and the enemy will be completely sheltered on reaching the foot of the hill. Not too high nor steep.

There should be no cover within easy range from which the enemy's infantry can pick off the gunners; as it is very difficult to shell troops out of a wood; and good rifle shots, posted under cover within say 800 to 1,000 yards of a battery, with unshaken nerves, and fairly sheltered from danger, should always silence the guns before the latter can silence them. Infantry would be destroyed attacking in the open, exposed to artillery fire; but they will avail themselves of woods, villages, and undulations No cover for enemy.

of ground; these, therefore, ought not to be within 1,000 yards of the position.

<small>Able to enfilade, but safe from that fire.</small>

The position should be such that some of the guns may enfilade part of the enemy's line, as a single successful enfilading shot creates more mischief and confusion than a dozen passing through the line direct.

But it should not be liable to enfilade. If the enemy takes advantage of this defect it will be fatal, and the guns will have to withdraw.

The front slope of the position should not be too gentle, as the effect of ricochet, shrapnel, and rifle fire will be increased. A short, steep slope immediately in front of the guns is preferable, affording a better view; the guns can fire safely over the infantry in both attack and defence; and when the enemy's troops are temporarily checked, the guns will gain time to retire.

<small>Command.</small>

To have slight command over the enemy's position is an advantage. But if the guns are in a lower position it will be difficult to shelter the ammunition waggons and limbers.

<small>Facility for retreat.</small>

There must be every facility for retreat, and no probability of the guns being cut off. Hence the site should not be unduly detached from the main position. There is risk in having the ground in rear intersected with banks, hedges, streams, or other obstacles. If these exist passages should be prepared for the guns.

The fire will be most effective when the slope of the distant ground is parallel to the trajectory. But such ground is often hidden, and ammunition cannot be wasted in firing at what is not seen.

Guns must be under cover as much as possible, and, to prevent the enemy observing the effects of his shell, it is well to have some trees, hedges, a field of corn, or even houses, 100 or 200 yards in front of the battery, as a mask, concealing the guns and yet not obstructing the view. Ground should be chosen for shelter to the horses and limbers while the guns are in action; but the latter must not be too far away for the supply of ammunition.

Cover for guns.

The guns should be posted a few yards from the edge of the plateau, or a little down the reverse slope, with the muzzles just pointing over the crest. The recoil may be decreased by the drag-shoe. It is possible to deceive the enemy as to the presence of a gun before it opens fire, by posting a troop of cavalry, or even the mounted detachment with horse artillery, in front of it.

Failing natural cover, resort must be had to artificial. Trenches may be dug 2 feet 6 inches deep close to the wheels and trails, in which the gunners stand when not actually serving their pieces. Spades are carried with the limbers, and sometimes gun-pits, (*vide* Fig. 109, Page 183), or a slight epaulment may be made. But as earth newly thrown up offers a conspicuous mark, it is well to connect it with shelter trenches for partial concealment.

Trenches and gun-pits.

A battery in action should not be immediately in front or in rear of its own troops, because it would impede their advance or retreat, offer a double mark to the enemy, and our own troops in front might be shaken.

Guns not to be just in front or in rear of other troops;

A stony position is bad, for the enemy's shells will burst with greatest effect, and fragments of stones

Nor on a stony position;

D 2

cause great annoyance. Marshy ground in front destroys the effect of ricochet fire, and even a ploughed field deadens it.

Nor near a conspicuous object. A battery should not be posted near any conspicuous land-mark, *e.g.* a prominent clump of trees, as the distance will probably have been calculated; and the background should be such as will not give the enemy a chance of seeing if his shells burst too long, and of correcting his aim.

A wall or a bank will cover guns from rifle fire; but if artillery fire opens against them, it is well to draw them back a short distance.

Besides line at full and close intervals, a battery moves in the following formations:—

Column of half-batteries and of divisions. * Column of half batteries, or of divisions, when each half battery or division is followed by its waggons (*vide* Fig. 29).

Of subdivisions. * Column of subdivisions, when the waggons are on the right or left of the guns (*vide* Fig. 30).

Of route. * Column of route, when each gun is followed in succession by its waggon.

Various kinds of fire. * There are various kinds of artillery fire, viz. with reference to the horizontal plane.

1. Frontal—directed perpendicularly or nearly so to the general line fired at.

2. Oblique—directed obliquely to the line fired at; it is more effective, because more searching than frontal.

* 3. Enfilade—raking the enemy's line, the guns being posted in prolongation of that line; it is most demoralising, there being no possibility of reply. It is usually offensive with reference to fortifications.

* 4. Flanking—directed along the front of, and

Fig. 29

COLUMN OF DIVISIONS.

110 yards

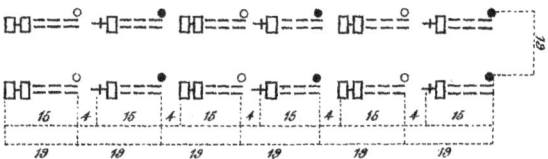

Fig. 30.

COLUMN OF SUBDIVISIONS.

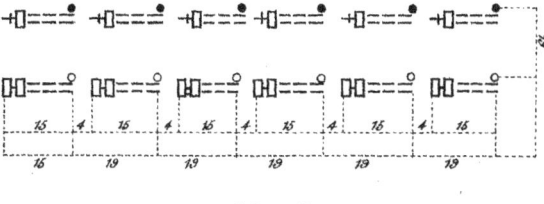

110 yards.

nearly parallel to the line to be defended, taking an enemy in flank, as he makes a direct attack. It has similar advantages to enfilade.

5. Cross fire—by which projectiles from guns posted in different parts of a position cross at a particular point of ground. It resembles flanking fire, and by its distracting and disconcerting action produces moral disorganisation. By means of this fire, rather than frontal fire, the long range of modern artillery is most effective.

And with reference to the vertical plane—

6. Direct fire—with service charges, at any angle of elevation not exceeding 15°.

7. Indirect or curved—with reduced charges, at any angle of elevation not exceeding 15°. It is used when the object fired at is unseen, *e.g.* at troops behind a hill or obstacle.

8. High angle fire—directed at greater elevation than 15°, with any charges, and it includes what was formerly known as vertical fire. It is so named from the general direction of the projectile on impact.

The projectiles carried by a battery are common shell, shrapnel shell, and case. Projectiles.

Common shell is employed against troops in mass or in column when shrapnel is not available; also against troops under cover, intrenchments, and buildings. It is a good range-finder when fired with a percussion fuze, for with a few trial shots, the ranges can be best estimated by the distinct puff of smoke on graze. Common shell.

It contains a large quantity of powder as a bursting charge, and therefore acts both as a missile and as a mine. It is most useful against earthworks; and its range is longer than the effective range of shrapnel. It is generally carried loaded. It answers well to shell troops out of villages, houses, woods, &c., and to set fire to buildings; but it is not to be compared to shrapnel against troops in the open.

There is little choice between common and shrapnel shell against men or guns under temporary cover. At Plevna the effect of each was inconsiderable physically; but it caused the enemy to lie close while the infantry advanced to the attack.

Plevna.

Shrapnel shell is employed against troops in the open in any formation, and to enfilade defensive lines. If used with skill at effective range, it gives far better results than common shell. Success depends on the striking velocity of the balls and splinters at rupture, correct timing, and the point of explosion. It is, therefore, usually fired with a time fuze, especially if range-finders are employed; for if percussion fuzes are used the graze reduces the velocity, and the balls fly upwards. But if the range is not known, percussion fuzes are first used.

Shrapnel shell.

The shells should be made to burst from 200 yards at short ranges, to 80 yards at long ranges short against troops in column; and from 100 yards to 50 yards short against troops in line; and they are most effective when they burst at 20 feet above plane at short, to 50 feet above plane at extreme range. The extreme range of shrapnel with the 16-pounder is 4,200 yards; with the 9-pounder 3,800 yards.

The effective fire of shrapnel shell at—

Effective range of shrapnel.

1,000 yards covers 300 to 400 yards depth of ground.
2,000 ,, 150 ,, 200 ,, ,,
2,500 ,, 100 ,, 150 ,, ,,

Its fire cannot be too direct against a deep object, but against a wide object with little depth, the more oblique the angle of fire is to the front the greater will be the area affected.

Time shrapnel is best against objects moving towards or away from the guns, with fuzes bored rather short against the former, and rather long against the latter.

Percussion shrapnel may be fairly effective against cavalry and artillery mounted, but the comparative effect of the shell with time and with percussion fuzes is as 7 : 1.

There are 63 bullets in the 9-pounder. The shell is usually carried loaded, and is distinguished from common shell by the head being painted red. In the Egyptian campaign it was used with great effect. But it is of little use when penetration is required. A shell reversed in a gun acts like case at very short ranges. *Employment in Egypt.*

Case shot consist of a tin cylinder filled with bullets. * *Case.*
The 9-pounder contains 110. It is effective up to 350 yards, or with an extra charge a little further, and is most useful against infantry and cavalry at very close quarters, to repel a sudden charge, as a parting discharge previous to a hand-to-hand fight with infantry, or to check a rush across a bridge, or through a gap or defile. At very close range, say 100 yards, double charges of case may be fired as a final blow, and there is no fear of bursting the gun. Case shot are kept handy for instant use, 2 in each axle-tree box.

Case must never be fired over the heads of one's own troops, though with care common shell and *Must never be fired over other troops.*

shrapnel may be so fired, provided the troops are a considerable distance in front.

Number of rounds per gun in the field.
The number of rounds carried by each gun with its waggon is as follows:—

	9-pounder.	16-pounder.	13-pounder.
Common shell ...	32	24	30
Shrapnel shell ...	112	72	108
Case	4	4	4
TOTAL ...	148	100	142

War establishment of batteries.
The war establishment of a—

Horse Artillery R.M.L. Battery is 182 officers and men.
Field 16-pounder is 201 ,,
Field 9-pounder is 175 ,,

Tactical subdivision of a battery.
A gun, with its waggon, men, horses, and stores, is called a subdivision. There are six subdivisions in a battery, numbered from 1 to 6; 2 subdivisions form a division, and the three divisions are termed right, centre, and left.

Gunners.
The gunners work and serve the gun, keep it, the ammunition, and stores in good order, and perform guard and escort duties and fatigues. They are armed with a sword-bayonet.

A few carbines are strapped on the carriages for use on guard, &c., but not for defence of the guns, lest the men neglect the proper duty of serving the guns.

Drivers.
The drivers ride, drive, and groom the horses, keep the harness clean, and mount stable guard at night.

The gun detachment of a field battery consists of the No. 1 and 8 gunners, marching with the guns or mounted on the carriages. In long marches they get an occasional lift.

Artillery cannot be moved about on its own ground, and for this purpose must have room both in front and on the flanks, which is the reason for intervals when acting with cavalry or infantry. The usual position of a battery is on the flank of a line. When forming up with other troops it should halt well in rear of the alignment, and as soon as the other arms are in position it will advance straight to its place.

Necessity for intervals.

Forming up with other troops.

Occasions sometimes happen when a battery has to pass through a line of infantry or cavalry. The artillery officer should then send word in time to their commander, who will make the necessary arrangements for letting the guns through.

Guns passing through a line.

On principle, the battery, being the unit, should not be broken up unless when absolutely necessary, *e.g.* certain cases of an advanced guard, and it is then recommended to break it up into divisions rather than into half-batteries.

Battery rarely to be broken up.

The present proportion of guns to men of the other arms is from 3 to 4 guns per 1,000, and this will probably be increased. Napoleon considered it ought to be 4 guns per thousand infantry and cavalry. In fighting a defensive battle, guns may be placed closer together than the regulation intervals, and sometimes there may be more than one tier of fire. 3 guns per 1,000, exclusive of the reserve troops, would seem to be generally sufficient. For a force encumbered with a numerous artillery is liable to certain disadvantages, viz. there may not be room on the position selected for all the guns; artillery cannot move quickly over broken ground, and can effect little without the other arms; and with long

Proportion of guns per 1000.

Limit to strength in artillery.

columns of guns and waggons confined to roads the infantry may be delayed, or some of the artillery will not be able to get to the front in time.

Range. *   Artillery may be employed effectively, without undue exposure to infantry fire, from 800 to 2,500 yards from the enemy; and if the ground and atmosphere are favourable, and range-finders available, guns may be used beyond 2,500 yards. In a special case, in order to gain some decisive advantage thereby, they may approach nearer than 800 yards, remaining exposed to infantry fire at the risk of loss of men and guns.

No artillery reserve in action.   As a rule, no guns should remain in reserve during an engagement; for they are seldom disabled by the enemy's fire, and can remain in action so long as there are supplies of men and ammunition to work them, and horses to change position when required.

Infantry. *   Now, as to infantry. It possesses this characteristic peculiar to itself, that it is the only arm which can act independently at all times, whether on the move or at rest, in attack or in defence. It can fight on every description of ground—on a plain, in mountains; in woods, villages, and towns; in an inclosed country and in the open.

Characteristics.

Functions. *   Its functions are to march, shoot, and fight well. It has to bear the whole brunt of war both on the march and in battle. It is self-protecting, and easy to train and equip; can move over any country, though much slower than the other arms. Infantry can fire on the move. But this does not mean that men may discharge their rifles while running. They must always halt to fire, although they may run some yards between the rounds, as may be ordered in the attack. Yet a body of infantry is incomplete alone, requiring engineers to make roads; cavalry to watch and to complete success; and artillery to prepare the way for and cover its advance, or protect its retreat. * Incomplete by itself.

Infantry is armed with a rifle and bayonet; and can apply "fire" and "shock" action individually or collectively, in combination or separately. Now-a-days its "fire action" is by far the most important, owing to the moral effect of superior fire. Nevertheless the bayonet should not be laid aside, but its use strongly inculcated. For this weapon gives the soldier a long reach, and an independence, especially against cavalry. Every drill practising the attack should conclude with the charge with fixed bayonets, except when there are two opposing forces. * Arms, and two modes of action.

* Importance of the bayonet.

Against savage warriors, who protect themselves with targets or shields, men should not thrust with the bayonet straight to their front, but in a slanting direction; each soldier pointing, not against his immediate opponent, but against the one opposite his right-hand comrade, so as to wound the enemy under the right arm before he can ward off the thrust. * Against men with shields.

The tactical unit of infantry is the battalion. Some consider the half-battalion, or the double company, as the fighting unit. But it is preferable to take the company as the fighting unit because once the battalion is disposed in extended order, causing loss of cohesion, owing to the superior fire-arms of the present day, no one officer can control * Tactical unit.

* Reason for this being the company.

it; and occasions arise for independent commands of lesser units, the battalion remaining the centre of action, and the general direction being maintained by its commander.

Pace. * The pace of infantry in quick time is 120 paces = 100 yards per minute; or 3 miles 720 yards an hour; in double time 165 paces = $151\frac{1}{4}$ yards per minute.

Frontage. * To find the front of a battalion in line, allow 2 feet for each man in the front rank, 9 guides, and 3 colour party.

Reducing files to paces. * To reduce files to paces, multiply by 8 and cut off the last figure.

The depth of a battalion in fours is equal to its front in line.

A battalion in column has the companies successively in rear of each other, and parallel at company or wheeling distance.

The depth of a battalion in column = the front in line − the front of the leading company + the depth of the rear company.

In a quarter column, the companies stand at 6 paces distance = 5 yards from guide to guide.

The depth of a battalion in quarter column = the number of companies × 5 − 2 yards. Thus, for 8 companies it will be 8 × 5 − 2 = 38 yards; for 6 companies, 6 × 5 − 2 = 28 yards.

30 paces = 25 yards is the interval left between battalions in line for elasticity and greater freedom of movement.

Front of a battalion in line. * Therefore the front of a battalion of 8 companies, of 100 men each, in line will be $800 + 24$ feet $= \dfrac{824}{3}$ = 275 yards (*vide* Fig. 31).

BATTALION IN LINE.

Fig. 31.

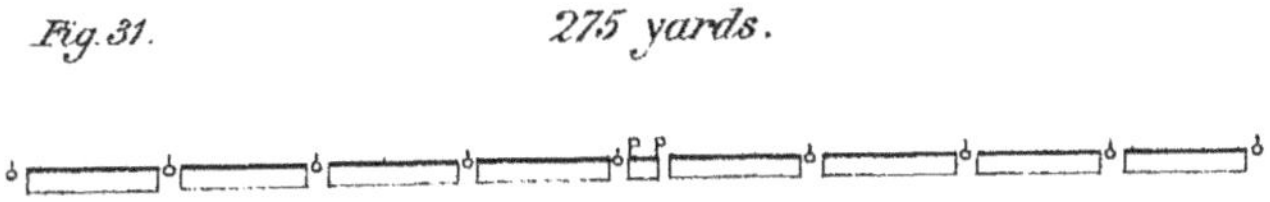
275 yards.

BATTALION IN COLUMN

Fig. 32.

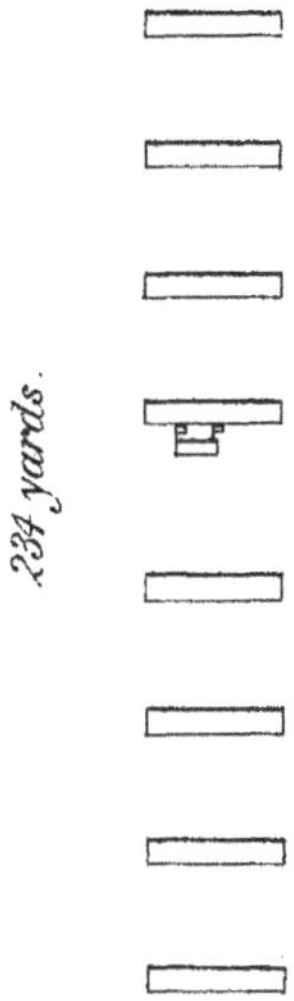
234 yards.

BATTALION IN QUARTER COLUMN

Fig 33.

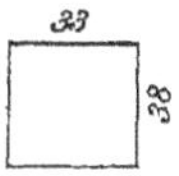
33
38

The front of a company of 50 files, extended at 4 paces interval, will be 200 paces = 167 yards. *Front of a company extended.

The depth of this battalion in fours will be the same as the front, 275 yards. *Fours.

The depth in column will be $824 - 102 + 9 = \dfrac{703}{3}$ *Column.

= 234 yards (*vide* Fig. 32).

The depth of this battalion in quarter column will be $8 \times 5 - 2 = 38$ yards (*vide* Fig. 33). *Quarter column.

The depth of infantry in line is about 3 yards from the front of the front rank men to the captains. *Depth of a line.

The detail of a company on a war footing is :—

Officers: 1 captain and 2 subalterns; 1 colour-sergeant, 4 sergeants, 1 lance-sergeant, 4 corporals, 3 lance-corporals, 110 privates, 1 piper, 2 buglers or drummers, 1 driver; total all ranks, 130. *War establishment of a company.

4 men should be instructed as stretcher-bearers, and 2 in army-signalling.

A battalion on a war establishment consists of:—

Officers	31
Non.-com. officers and men . .	1,044
Drivers for regimental transport .	22
Total officers and men .	1,097
Riding horses	12
Draught horses	44
Total horses	56

*War establishment of a battalion.

Infantry, in its attack, endeavours to destroy, capture, or drive the enemy from his position. For this, superior fire alone is rarely sufficient; therefore, to ensure success we must be ready at the proper moment to deliver the charge, and the charge alone *Infantry in attack.

* will still more rarely enable infantry to gain its end; hence a combination of these two is necessary to ensure victory.

<small>On the defensive.</small> * Infantry on the defensive, firing from behind cover, must strive to demoralise and practically destroy the enemy; and that done, full advantage must be <small>Waterloo.</small> secured by delivering a counterstroke, as at Waterloo, combining offensive fire with the charge.

Instances, however, occur when defensive fire alone is permissible, as when supporting artillery. The object then is to keep off the enemy's riflemen, and save the guns from capture. Or when the defending <small>Plevna.</small> * troops are greatly outnumbered, as at Plevna.

<small>Distinction between "extended order" and "skirmishing."</small> * Infantry is extended for purposes of attack and for skirmishing. But these must not be confounded. Men are formed in "extended order," properly so-called, to bring them up gradually with as little loss as possible over the dangerous ground to a point where the hottest fire can be directed against the enemy's position, and then the line is reformed compactly to deliver the final assault. Whereas the duty of skirmishers is to cover a body of troops behind them, and to feel for an enemy in an inclosed or wooded country. When the enemy is discovered, the skirmishers double round the flanks and through the intervals, clearing the front, and reform. Or else they lie down till the other troops pass over <small>A battalion skirmishing.</small> * them. Sometimes a whole battalion is employed to skirmish in front of a division, and when the object is accomplished, it will reform and join the rear brigade. Greater latitude is allowed in skirmishing, and the number of paces for the extension will be named. But in "extended order," the interval

between files is 4 paces or less, including the space occupied by the men.

Although men in "extended order" are individually more independent than in "close order," yet "extended" order requires of them more self-confidence, and the strictest discipline. They must, moreover, be alert and nimble of foot. The men of a company should frequently look for the signals or direction of the captain; those of a section must attend to their section-leader. All words of command must be passed by supernumeraries. Section commanders should know the number of files in their respective sections, and their right and left hand men. Their post is a few paces in rear of the centre of their sections, having their eyes on all the men.

* Self-confidence and strictest discipline essential in extended order.

CHAPTER IV.

INFANTRY IN ATTACK AND UNDER OTHER CIRCUMSTANCES.

Necessity for developing the soldier's warlike character. * BESIDES steady drill on the barrack square, and the perfecting of the men's training, efforts must be made to secure a further advance in the province of their intellect, and in developing their warlike character, especially in extended order; for close formations are not so much employed in action, being suitable only for bringing up troops for the fight, although they may be used in a few cases by night, or in surprises, or against cavalry. And troops held in 2nd and 3rd lines are in close formation. The nearest formed body of troops behind the "fighting line" is called the support. The company or file by which the others march is called the directing company or file.

Directing company or file. *

Duties of a section commander. * The duties of a section commander, moving to the attack, are:—to see that the men use the best cover for concealing their movements, and for inflicting loss on the enemy; to decide what object is to be fired at, the distance and sighting of the rifles, and the kind of fire; to keep their men in correct formation, and to lead them properly to the front and rear; to control the expenditure of ammunition. This last is most important, and a section commander should name certain men to fire, or for drill purposes, direct the front and rear ranks to fire alternately, or right files and then left files. A better plan still is to order the section to fire a certain number of rounds—say from 3 to 5—after the expenditure of which there will be a pause in the firing, to allow the smoke to clear off, to subdue any excitement or unsteadiness in the men, and to give them fresh orders.

On a favourable opportunity occurring, he may seize a position by the combined movement of his section, provided he will not expose the men to needless loss, and support is at hand.

A battalion is subdivided for attack into the fighting line, supports, and main body. [* Subdivision of a battalion in the attack.]

About 4 marksmen per company, under an officer, are selected as scouts, moving in advance of the fighting line, until the enemy's strength is ascertained, which will probably be about 400 to 500 yards from his position, when they should lie down, and await the arrival of the fighting line. [* Scouts.]

Scouts, connecting files, and all others, make use of the code of signals referred to under the head of advanced guards, as occasion may require. (*Vide* pp. 71, 72.) [* Signals to be used.]

Bugle sounds may disclose our intentions. The commanding officer's bugle is generally sufficient. It is only an extended line that moves by bugle sound; the supports and main body move by word of command. The advance must be made as quietly as possible, captains and supernumeraries passing all words of command. [* In preference to bugle sounds.]

E

The commanding officer. *	When a battalion is ordered to attack, the commanding officer points out to the captains of Nos. 1 and 3 companies the enemy's position, and selects some conspicuous object for the centre file to march on. Whenever practicable, the extension should be
When possible makes all dispositions under cover. *	made and everything ready for the advance under cover; then at a signal, or by word, the front line moves on, followed by the others at the regular distances. (*Vide* Fig. 34, representing the 1st stage.)
First stage of attack. *	The men are to avail themselves of the slightest inequality of ground, as cover, avoiding all needless exposure. But they must not seek cover, if by so doing they lose their intervals, and fail to move direct to their front. Should artillery fire open upon them, they must not seek cover except by word of command. About 800 to 1000 yards from the position, according to what may be seen of the enemy, the fighting line will open fire, the rear rank coming up
The two men of each file to keep together. *	on the left of the front rank, so as to form rank entire, but the two men of each file keeping together. If the bugle sounds "the fire," this is permissive only, as it is useless to waste any rounds when there is no good object seen, or the fire would not be effective.
Fighting line. *	The fighting line and the supports are under the officer second in command. He conducts the advance either in a general line or by alternate companies until within 600 yards of the position (*vide* Fig. 35,
The second stage. *	the 2nd stage), when the fire should begin to have good effect; the advance being continued by short rushes of 30 or 40 yards from cover to cover, the men lying down to fire, protecting their comrades on the move. The greatest care is required in practising this, so that the men do not close in or open out;

A BATTALION ATTACKING.

1ST STAGE.
Fig. 34.

Say 700 yards from enemy's position.

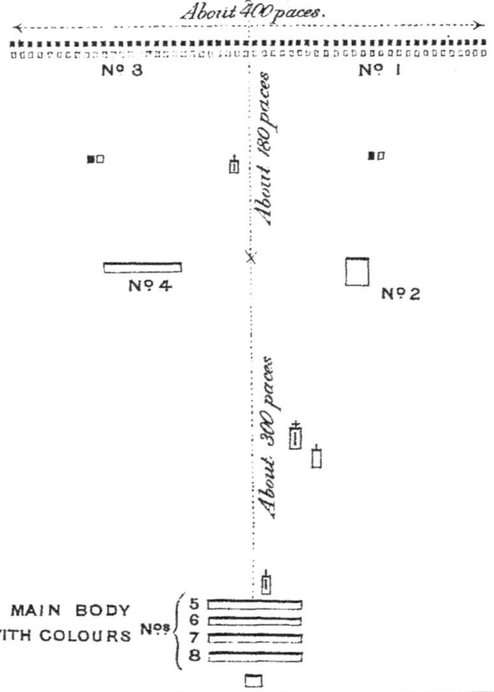

About 400 paces.

No. 1

No. 2

No. 3

About 180 paces

No. 4

About 300 paces

MAIN BODY WITH COLOURS Nos. 5 6 7 8

IIND STAGE.
Fig. 35.

500 to 600 paces from enemy's position

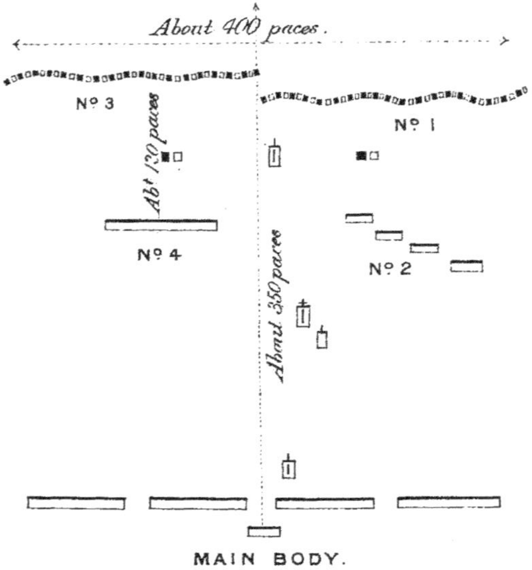

About 400 paces.

No. 1

No. 2

No. 3

Abt 120 paces

No. 4

About 350 paces

MAIN BODY.

Fighting line firing in rank entire, advancing by alternate companies, supports in two ranks moving as convenient. Main body deployed with intervals between companies.

Fig 36.

A BATTALION ATTACKING.

IIIRD STAGE.

From 300 to 400 paces from enemy's position.

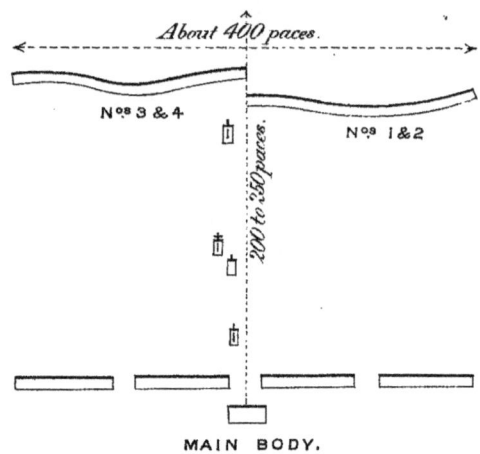

Fighting line reinforced by supports firing in rank entire, advancing by alternate combined companies.

Fig 37.

IVTH STAGE.

About 150 paces from enemy's position.

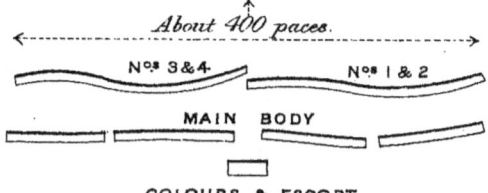

Fighting line checked. Main body reinforced according to circumstances either in rank entire, or in two ranks.

that they double steadily, and do not mask the fire of those lying down. As a rule, the alternate bodies, moving up, will not go beyond those in front, but lie down in line with them.

Nos. 2 and 4 companies form the supports, following the extended companies at about 150 yards; but gradually decreasing that distance. They are intended to make good losses in, and to reinforce the fighting line when necessary, so that its fire may never slacken; also to protect its flanks, being ever ready to pour a heavy fire upon any of the enemy's cavalry or infantry that may be threatening; and supports give confidence to the fighting line by having a compact body of men of the same battalion close at hand. Every supporting company sends on a connecting file to communicate with the company in its front.

* Supports

The supports adopt any formation suitable to the ground. In the open they move at open files, at 1 pace interval. They may form column, for better shelter, behind a house or hillock; and if on an exposed flank, échelon of sections is advisable.

When the fighting line experiences difficulty from the enemy's fire, it will be reinforced. This is at the discretion of the field officer leading the advance, who should remember that the nearer the attacking line can push on without reinforcement, the greater will be the effect when it does take place.

* Reinforcing discretional.

Each support should extend so as just to occupy the front of the company reinforced. At drill the files should be directed on the left of the corresponding files in front. This is the 3rd stage (*vide* Fig. 36), and the advance is continued by rushes by alternate

* Extension of supports.

Third stage.

combined companies of about 30 yards, at a smart, but steady double, by word of command of the senior officer of each combined company, the men lying down keeping up the fire. When the men who advanced are settled on the ground, they open fire, and the rear combined company is brought up into line, and so on. The utmost care must be taken to preserve the direction and proper intervals. No firing is permitted except when halted.

Reserve. * The main body or reserve is to reinforce, and otherwise to assist the companies in front, so as to secure the enemy's position by a final rush. It may be in column by the left, or in any formation the commanding officer thinks best. When deployed, intervals between companies may vary from 3 to 20

Use of fours in rough ground. * paces. In crossing rough ground, intersected with obstacles, it is desirable to advance in fours from the flanks of companies, which will front form when cleared. As a rule, under fire, the main body of a central battalion will move in line with files at one

Echelon on a flank. * pace interval; that of a flank battalion should move in échelon.

About 200 yards from the enemy, the commanding officer, having resolved to force the position by a front attack, reinforces with the reserve (*vide*

Fourth stage. * Fig. 37), the 4th stage. The companies, advancing in line, form rank entire, fix bayonets on the march,

Reinforcement by reserve. * and slope arms. When halted they form a rear rank to the men in the fighting line, who will have fixed bayonets, as soon as they see the main body coming up. The line may be ordered to advance, and halt to pour in the hottest fire for a few moments, then to advance again. When within charging distance, the signal or order to charge will be given; the buglers, drummers and pipers sound the charge; the men quicken their pace and cheer.

<small>* The charge.</small>

If guns or cavalry are ordered to pass through an extended line, in which there are no intervals, sufficient files must close right and left to let them through, and at once resume their places in line.

<small>* Opening for cavalry and guns to pass.</small>

In practising the attack with a company, the enemy's position should be marked, and some lateral limits fixed, outside which the movements are not to take place. For it is necessary that the men are instructed as if they really formed part of a more extended line.

<small>* How to practise the attack.</small>

A skeleton enemy is most useful. Thus four drummer boys may be posted on the ground to be advanced over; two representing parties of infantry, and the others, with camp colours, parties of cavalry, who should endeavour to work round the flanks, and raise the flags to show they are about to charge.

<small>* Skeleton enemy.</small>

Although rarely the same troops, who have stormed a position, will be able to continue a further advance, pursuit being left to the second line passing over them; yet sometimes it may occur that the enemy has withdrawn after the fighting line has been reinforced, and it is required to continue the advance in the original formation. In such a case, the men last brought up will form the new front line; and on the order being given, the captains, or section commanders, calling the attention of their men, will direct their further advance; the remainder, having closed, follow at the proper distances as supports, or reserve.

<small>Conduct of a further advance.</small>

When the battalion is much mixed, the quickest

Reforming the battalion. *	way to re-form is to call out the markers and sound the assembly.

To fix bayonets on the march, or lying down, the method is the same as that laid down in page 56 *Rifle Exercises.*

Fixing and unfixing bayonets on the march. *	To unfix bayonets on the march, the rifle should be carried in the left hand at the trail, barrel slanting upwards, the bayonet is unfixed with the right hand, the rifle changed to the right side, and carried below the right arm-pit, while returning the bayonet to the scabbard.
Ammunition carried by the soldier. *	Every infantry soldier carries 70 rounds of ammunition in action. Three ammunition carts per battalion follow, each carrying 9,600 rounds. Of these
Ammunition carts. *	one cart accompanies the battalion into action, about 20 yards in rear of the main body, until approaching the fighting line, when it must be placed under the nearest cover, ready to furnish the supply; the other two are parked in rear, with those of the other battalions of the brigade, awaiting orders. Each company of the fighting line should have its ammunition served out to it previous to an advance;
Replenishing the pouches from the regimental cart. *	and a careful system should be adopted for replenishing the pouches from the regimental cart. Pioneers or bandsmen are generally employed for this duty. They ought to be provided with a double bag worn over the shoulders. They must know how to get from their companies to the ammunition cart. The ammunition on the dead and wounded should be collected as far as possible.
Disposition of a battalion for defence. *	When a battalion is ordered to occupy a defensive position, the disposition is much the same as for attack. The fighting line is made as strong as a free

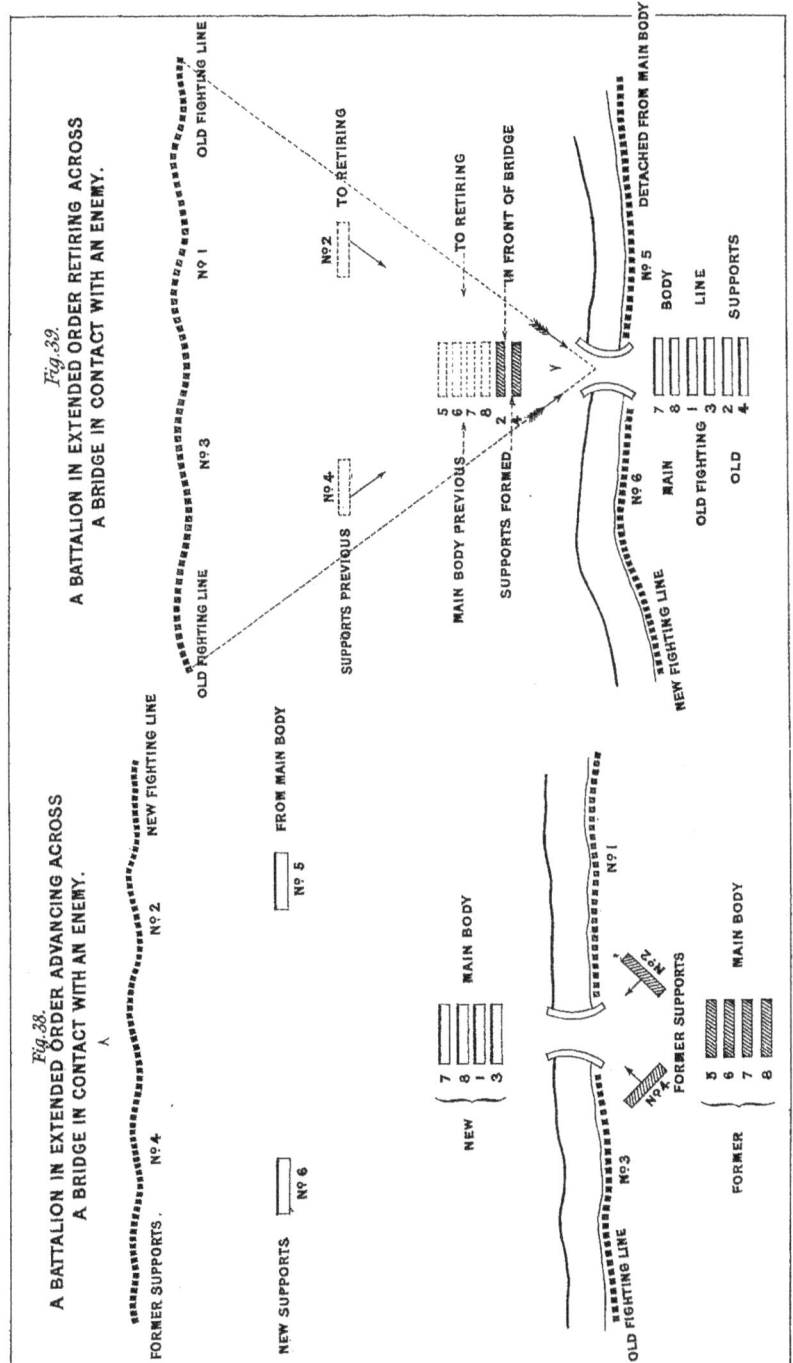

use of the rifle will admit. It is, therefore, usual to place 3 companies in front line instead of 2; and these open fire as soon as the enemy comes within effective musketry range. Ranges of several objects should be taken previously. The supports are posted under cover, close by, to replace casualties in the fighting line. The main body is placed centrally, so that when the enemy's attack is developed, part of it, or the whole, may be moved to a convenient spot to deliver a counter attack. * Ranges to be taken.

The object of the defence being to hold the position, the enemy must be kept at a distance by fire. But should he, notwithstanding, attempt to charge, the counter attack must repel him with fire and the bayonet. * Counter attack.

When a battalion is obliged to retire, the retreat should be conducted by companies to positions chosen for mutual support; or by alternate half-battalions, each under its own field officer, retires 100 or 150 yards, halts, fronts, and opens fire, covering the retreat of the other. * Battalion in retreat.

In order to pass a defile, or advance across a bridge in presence of an enemy, the principle is understood by reference to figure 38, taken from the *Field Exercises*. On reaching the defile, river, or canal, the fighting line opens fire; the supports close on the one nearest the defile or bridge, and with the main body, if necessary, force the passage with the bayonet. This done, the supports extend. The main body retains possession of the defile or bridge. The old fighting line continues firing until its front is masked, and then re-forms in rear of the reserve, and the advance is continued. * Advancing across a bridge in presence of an enemy.

Retiring across a bridge in presence of an enemy. * To retire across a bridge, or through a defile (*vide* Fig. 39), the plan is for the main body to pass first, and extend 2 companies, lining the river bank on either side. These open fire when the front is clear. The supports close, and halt in front of the bridge until the withdrawal of the fighting line, the men of which incline towards the bridge, run quickly over, and form in rear of the reserve. The supports cross last, and the whole will be prepared to defend the passage or retire.

To practise these two movements on parade a bridge may be marked with camp colours.

Separate bridges for different arms. * When a large force has to pass a river, separate bridges should, if possible, be provided for infantry, cavalry, and artillery. The several arms should not be mixed.

Precautions in crossing bridges. * In crossing, infantry must break step, and the band stop playing. Files and sections are not to close up. Cavalry, as a rule, cross in file at a walk. Artillery, with carriages up to guns of position, 40 pounders, must cross fully horsed. All halting on the bridge should be avoided.

Conduct of the soldier in action. * In action the soldier must move in a free and unconstrained manner, and at a smart pace. Silence must be enforced. The whole attention must be paid to the section commander, and that of the latter to the captain. Arms are carried at the "trail"; in wet weather at the "secure." If required to double, he will be specially ordered to do so by signal or word of command; and then there must be no running, but each file must never lose connection with those on his right and left, specially observing the direction of the advance and conforming to it. Each rear-rank man must never separate from his front-rank man.

Men or files, when extended, may be ordered to execute independent firing—so many rounds; or to fire volleys in extended order; or, if closed, volleys in close formation; and independent firing in close formation. * Different kinds of fire.

In volley firing, every man firing before the others should be punished.

There are short-range firing, viz. at 400 yards and under, and long-range firing, viz. from 400 to 800 yards, beyond which distance firing is only ordered under special circumstances. Every man must carefully husband his ammunition; for it is very easy to come to the end of it.

All firing must at once cease at the word "Cease firing"; or at a long-drawn whistle sound from the captain. This sound also means "Pay attention." * Signal for cease firing.

The "alert" is the only bugle sound to be used in action when cavalry threaten. * The "alert" the only bugle sound.

Each soldier must be posted in such a manner as to have a clear field of fire. He must learn how to make good use of ground both as cover and as a rest for his rifle—*e.g.*, undulations of the surface, walls, hedges, large and small trees; and further, how to fire from loopholes, windows, shelter trenches, and earthworks. The enemy must be "stalked," and his position approached as close as possible by the soldier unobserved under cover. In small engagements this may sometimes be done by creeping forward. He must be a good and rapid judger of distance. * Soldiers must learn the use of ground; * and judge distance.

Sometimes the fight is stationary, the men on

both sides remaining under cover, firing at each
*other for a considerable time. Lastly, we may consider soldiers properly trained, when, in changing position, they are able to take up their places in the new position in such a manner as to be well under cover, and favourably posted for firing with great rapidity, yet with steadiness, and without crowding or unduly opening out.

*The three military positions authorised for firing in the *Rifle Exercises* are—standing, kneeling, and lying down, which last is the most usual in action. When the line approaches the enemy's position, and the fight is raging, so that the section commanders
*cannot name the range, it is well for the men to use the 200 yards' sight from 300 yards and under, aiming rather above the object up to 200 yards, and a little below it beyond 200 yards.

*In an engagement between infantry there is a continuous musketry fire. Each man must make every effort to remain cool, notwithstanding the striking of bullets and the noise. He should always take a
*distinct aim, and never fire at random. In independent firing aim must be taken quickly.

The object to fire at is the enemy's fighting line. If nothing is visible but smoke then the thickest layer should be aimed at. Always fire at the officers and mounted men, and above all, at columns. At night, companies will be in close formation, and must be ready to fire volleys or deliver a bayonet charge.

In attacking artillery, infantry should fire long-range volleys, or what is better still, gain, if possible, a point within short range of a flank of the battery, by means of such cover as folds of ground, bushes,

Sidenotes:
When considered properly trained.
Three military positions for firing.
Adjusting the sights at close range.
Infantry against infantry.
Never fire at random.
Infantry against artillery.

standing corn, from which the men can open independent firing. The battery should then be charged. If it is moving, or about to limber up, fire at the horses; if it is in action fire at the line of guns itself. Rapid advances and diagonal movements are the best protection against artillery fire. If the battery is captured and has to be abandoned, spike the guns if they are muzzle-loaders, or take away the removable part of the breech action if they are breech-loaders. * Fire at the horses moving, at the line of guns in action.

Provided we remain cool, we possess a decided superiority over cavalry in action, both in single combat and in bodies, cavalry charge in échelon, *i.e.* in detachments following each other in rapid succession. What has to be done is to fire steadily at short ranges, with the sights of the rifles at 400 yards, aiming at the foot of the object, the advancing cavalry will then be struck throughout that distance. A fighting line favourably posted should not close to receive cavalry, but the men should retain their position, directing their fire against the horsemen. But if the cavalry approach within 50 yards the fighting line should stand up. This will startle both man and horse, and it is easier to aim standing at a mounted man than when lying down. * Good infantry invincible by cavalry.

* Adjustment of sights.

* As a rule avoid squares;

Men extending, lying down in the open, unprotected by any cover, if not under a heavy fire, should close and form sections or company. * Unless unprotected in the open.

When cavalry appears on the flanks these must be thrown back, when it appears in rear the rear rank should be turned about. Volleys are best against cavalry.

In the event of all the ammunition being expended, squares and rallying squares must be formed to resist with the bayonet. If cavalry really ride the men down they should throw themselves on the ground, but jump up again immediately, form either in line or rallying square, and fire after the cavalry, or at the échelons coming up. No soldier should surrender or allow himself to be taken prisoner.

Single combat. * In single combat an infantry soldier may allow a mounted man to come within 30 yards before he fires. If he misses he should fire again. If he has no more ammunition he should jump aside as the cavalry soldier charges, and thrust at his horse.

Formations to resist cavalry. * The chief formations to resist cavalry are squares and line. Steady, well-disciplined infantry have little to fear.

The "thin red line." * The "thin red line" receiving the masses of the Russian cavalry at Balaklava is a conspicuous instance. Company squares are formed by an extended line and the supports when threatened by a very large body of cavalry. Care must be taken that the squares are in échelon.

Rallying squares. * Rallying squares are quickly made, when an officer for each square holds up his sword, turning towards the enemy as a rallying point, and the men at the command run towards the officer, form round him without crowding, fix bayonets or swords, and turn outwards.

Closing on flanks of section. * An excellent plan, easy of application, is for the men of each section to close rapidly on the right or left file of that section, according to which flank is threatened (*vide* Fig. 40), the flank file making a half turn outwards; or if both flanks are threatened

Fig. 40.
SECTIONS RECEIVING CAVALRY.

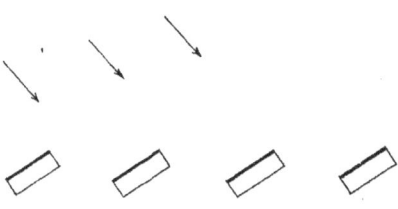

Fig. 41.
ESCORTS FOR ARTILLERY.

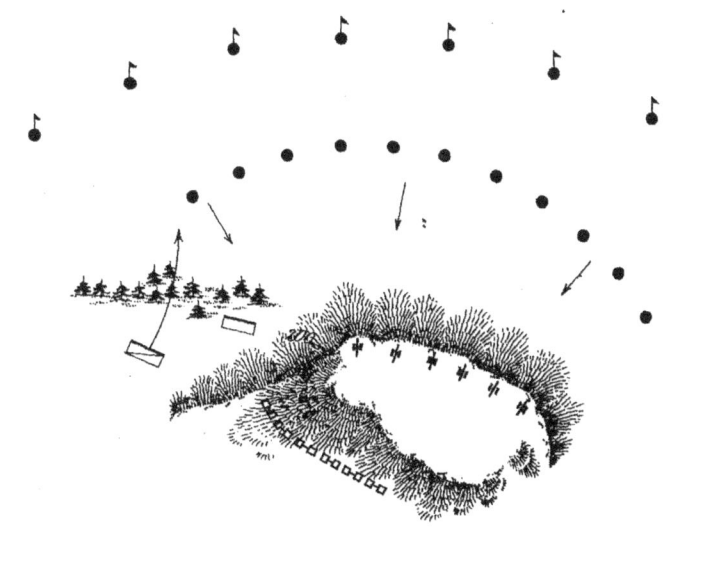

the two right sections of a company will close on their right files and the two left sections on the left. The word of command for drill will be—"From the right" (or left), or, "from both flanks prepare for cavalry." The men fix bayonets, the front rank drop on the knee, and the rear rank open fire by command of the section leaders. Care must be taken that the sections are not wheeled up too much.

As a rule they will be formed in échelon, and able to fire clear of each other. If not, they must move into échelon. * Echelon.

A support, in open files, and hard pressed, may close and form an oval round its captain, or the usual company square, closing on the centre. The necessity for the main body to form square is exceptional. * Company squares.

The great disadvantage of squares is the large mark they offer to artillery; and often cavalry will charge the flanks of infantry with that very object. * Disadvantage of squares.

A company extended, on the approach of cavalry, may also form groups, each composed of a right and left file, or, if the company has been reinforced, the two files next each other will run together, and stand back to back. * Groups.

In resisting cavalry, much depends upon the nature of the ground on which the infantry are, and of that over which the cavalry has to pass; also the quality of that cavalry, and the morale and cohesion possessed by the infantry.

As before mentioned, a battalion may receive cavalry in line. But there must be no inequalities of ground to protect the cavalry, or enable it to form unperceived anywhere near. Often a flank company is thrown back, and another forward; or both may be thrown back.

Battalion square. * A battalion square is formed from line, column, quarter-column, or column of double companies. Whenever it is possible for cavalry to approach unperceived, infantry must be ready to form square two-deep, or four-deep, at any moment.

Infantry in a siege. * It may happen that infantry has to remain rifle in hand in the trenches for days together in presence of an enemy. Each soldier should be provided with about 200 rounds, and special rations. It may be required to search a particular work in the enemy's lines with long range fire, aim must then be taken at the crest of the parapet, which usually stands out clearly, and particularly at the embrasures.

Conduct of an assault. * If a work has to be taken by assault, precise directions are issued beforehand as to the conduct of the different portions of the attacking force, and as to their direction. Success is insured only by the most reckless bravery, and the most expeditious crossing of the ditch. On reaching the berm, the men halt for a moment to form, then mount the parapet closed up as much as possible.

Instructions to the German infantry soldier in attack. * A small pamphlet has been published for $1\frac{1}{2}d.$, by a German officer of great experience in war, and a well-known writer, containing invaluable rules for the conduct of the men in action, and in the various incidents of the same. The following are some extracts therefrom :—

When the whistling of bullets commences, the men must constantly remember to show no signs of weakness, even in the most perilous moments of the fight.

It is strictly prohibited to attend to the dead or wounded. It is better not to let the eyes rest on them. Men who are slightly wounded are not allowed to withdraw from the fight.

Not to leave the ranks to attend the wounded.

Wounded men coming out of the fight, and returning to the rear, must not dare to dishearten the reinforcing troops. They must, on the contrary, put up with pain, and strive to groan as little as possible.

Throwing packs away without orders will be severely punished.

Every soldier must strive to remain with his company or section during the fight, as he cannot, single-handed, effect anything of importance. His concentrated attention must, therefore, be fixed on his leader.

Should a soldier, through no fault of his own, become separated from his company he will join the nearest fighting body of troops, and place himself under the orders of the officer or N.C. officer in command of it, to whom he will render the implicit obedience due to his own immediate superiors.

Duty of the soldier when separated from his company.

Whenever bodies of troops become intermingled in action (which frequently is unavoidable), the superior officer who assumes command is to be obeyed without hesitation.

When troops are mixed the senior officer commands.

The soldier must bear in mind that long-range firing rarely does much execution, and that its employment remains as matter of uncertainty; the leaders are responsible for the results obtained thereby. Consequently, during an attack, no firing must on any account take place at greater distances than 400 yards, unless expressly ordered.

Each man must adjust his sights independently.
The soldier must, above all, remember to adjust his sights correctly. This is particularly necessary, as the company and section leaders are frequently put out of action, and the men are then thrown on their own resources.

Should the soldier observe any movements or changes in the enemy's fighting line that may appear to him of importance, he must report them to his superior.

Every soldier must be able to convey an order clearly, and to make a clear report in the midst of the fight.

During the advance no man is allowed to hesitate or halt, no matter how heavy the enemy's fire may be, or how severe the losses suffered. It is only when the officer gives the command, "Halt," that a halt is made, and a position is immediately taken up.

In closing on the enemy at the run, the advance must be continued without a check right up to the enemy's position. Should this not be done, and should the assailants run back, they are as good as dead, as they have to recross the ground under a murderous fire.

Effect of a determined charge.
A determined charge really driven right home will invariably be successful.

If, during the fight, the enemy makes a determined forward movement, the principal thing is not to allow oneself to be intimidated. Should some men lose courage and run away, they must be immediately brought back and encouraged by their comrades.

In the moment of danger the soldier must look to his leader, for he it is who orders what has to be done.

<small>In difficulty look to the leader;</small>

The enemy's attack must either be met with a withering fire, or he must be immediately charged.

Should the firing be so heavy that the orders of the officers cannot be understood, attention must be paid to their signals with the sword, and to their personal movements.

<small>and attend to signals.</small>

When the enemy has been driven out of a position, the individual soldier must on no account rush on in pursuit, but wait the orders of his leader, and in the meantime continue firing on the enemy.

The men must be able rapidly to re-assemble after such an attack.

On the defensive the men must be determined not to move from the position they are to hold.

<small>On the defensive.</small>

The firing is commenced either by order or by signal. The soldier should not allow himself to be intimidated by the enemy's shouts and close approach. Every man should remain lying down, and fire steadily. A position should not be evacuated except by a distinct command.

The enemy will not usually be able to withstand the firing of such a determined body of troops, but will run back. Should he, however, really come to close quarters, the men must be determined to engage him even hand to hand.

With equally good arms victory will generally fall to the combatant possessing the greatest courage and coolness and the best discipline.

F

When casualties occur, lance corporals and best soldiers must replace their seniors.

* When the officers and N. C. officers have fallen, every honourable soldier must strive to replace them. The lance corporals, the oldest and ablest men, take command and lead their comrades.

Should unfortunate circumstances and the decided superiority of the enemy compel a body of troops to fall back, no soldier must appear discouraged, or retire further than to the spot where the leader orders a halt, or to the pre-arranged place of assembly.

Should a soldier find himself separated from his corps after an action, he will proceed in search of it and join it without delay. All aimless roaming about on the battle-field will be severely punished.

Should a soldier take part in the fight in the ranks of another corps, he must carefully remember which it was, for the purpose of producing witnesses as to his conduct.

To be humane and not to plunder.

* The soldier must behave in a noble and humane manner towards the enemy or prisoners. Under no circumstances whatever should their private property be taken from them, and they should never be ill-used without necessity.

The white flag with a red cross denotes dressing stations and hospitals; a white band with a red cross worn on the arm denotes medical officers, bearers, &c. These, in accordance with the Geneva convention, are not to be fired on.

Firing off a *feu-de-joie* after the battle, plundering captured baggage, or any similar misdemeanour, is strictly prohibited.

Furthermore, the soldier should be ready after an

action, notwithstanding his exhaustion, to start off in pursuit of the enemy, should this be ordered, for it is only by a rapid and active pursuit that the victory is completed, and further fighting avoided.

* To be ready for pursuit.

Escorts for Artillery.

Artillery is helpless on the move. Occasions arise when a battery is sent forward to an advanced position for a special purpose, or detached to a flank, exposed to the enemy's infantry. Then if the fire of the battery is diverted in self-defence, the object for which it took position cannot be attained. Hence it requires other troops in support.

Necessity for an escort.

Guns in action have to be guarded against a sudden flank attack of cavalry, combined with a converging attack in front by skirmishers; and, also, against the enemy's infantry stealthily advancing in extended order under cover in front, or on a flank.

Cavalry or infantry cannot be taken from their legitimate work to form a permanent escort, so the proper course is for the battery commander to apply for a squadron of cavalry to accompany his guns when detached, as protection on the march, and when first coming into action; and for a company of infantry to follow rapidly to replace the cavalry.

When required how to be applied for.

If the officer commanding the escort chance to be senior to the artillery officer, he should in no way interfere, but assist the latter in every way, by defending the guns, and guarding against surprise.

Duty of the officer commanding the escort.

The escort should be formed as an advanced guard in a forward march, and as a rear guard during a

* Formation on the march.

retreat, guarding specially an exposed flank, or part may be at the head and part at the rear of the battery. When marching by a road, all adjacent roads must be patrolled.

Disposition of an escort of cavalry. * When the cavalry escort arrives at the position, if an attack by cavalry is expected, scouts must be sent well on to the front and flanks for timely warning, and the remainder take post some 200 yards on the exposed flank of the battery in échelon, ready to charge. The best disposition against infantry is to dismount and extend half the escort, retaining the rest in support. But the infantry should hurry up and relieve them.

Of an infantry escort. * On arrival, the infantry escort should extend half the men at wide intervals to watch and give the alarm, much like outposts. They should be so posted as to keep the enemy's skirmishers at a distance of 1,000 yards at least, and to prevent anything coming within a mile of the battery unseen. As far as possible, the enemy must be prevented from occupying any cover within range. The remainder of the escort should form the support 150 or 200 yards from and in line with the battery on its exposed flank, and 100 yards away from the waggons, for fear of an accidental explosion.

The escort must move on the flanks of the guns so as not to mask their fire, unless they are in action on such high ground as to be able to fire over their heads; and possibly, even the escort may retire under the very guns themselves (*vide* Fig. 41).

Infantry escort may be conveyed on artillery waggons. * It is possible to mount infantry on the waggons. A horse artillery battery can take a company, and a field battery about half a company.

CHAPTER V.

ADVANCED GUARDS, FLANKING PARTIES, REAR GUARDS.

An Advanced Guard.

INSTRUCTIONS are to be found in Part VI. *Field Exercises*. An advanced guard furnishes the "eyes and ears" of a force on the march; an army unprovided with one is liable to come unexpectedly in contact with the enemy, and even if the opposing force be small, the army must halt to disperse it. Therefore, whether the advance be made on one or more roads, each column must be preceded by an advanced guard, to cover and conceal the march and formation of the main body; to feel the way through the country, searching well in front and on the flanks of the enemy's line of march; to give timely notice of his vicinity or approach, and keep him in check until the main body has time to prepare for action; also, to remove obstacles, prevent delay, facilitate the march, and guard against surprise. *Necessity for an advanced guard.*

In some cases it is sent far in advance to seize a post, or to anticipate the enemy on some important point. *Sometimes sent far in advance.*

General duties.	* An advanced guard should not compromise itself with a superior force, but must not let the march be delayed by insignificant demonstrations. It must either attack promptly, or fall back slowly disputing the ground. Occasionally its duty is to engage and hold fast the enemy, and for this purpose it is strengthened.
Specially organised, usually of the three arms.	* It is a specially organised force, consisting, as a rule, of all arms; having more cavalry in an open, and more infantry in an inclosed or hilly country, with artillery to shell the enemy out of farms and sheltered ground.
On what depends its composition and strength.	The composition and strength depend upon the distance in front of the main body, the object in view, the nature of the country, and other circumstances, appreciated only on the spot.
When cavalry screen precedes the army.	If the army is preceded by the cavalry screen, and the enemy is at a distance, the advanced guard need only be of sufficient strength for observation. But, approaching the enemy, the cavalry screen will encounter his infantry, and the advanced guard must be ready to furnish support.
When there is no cavalry screen.	If there is no cavalry screen, it must be tactically complete — a miniature army — supplemented by cavalry to obtain information, and protect the column in rear, by extending the leading patrols sufficiently on each side.
Infantry on the flanks in mountains. Proportion to the army covered.	* In a mountainous district parties of infantry should move along the heights on each flank. The larger the army the greater the number of men which can be spared for the advanced guard. In a forward march $\frac{1}{8}$ the effective strength of the main body, or $\frac{2}{3}$ the strength of all the detached

parties surrounding the army is an average proportion.

But in 1870 the Germans sometimes employed $\frac{1}{4}$ to $\frac{1}{2}$. A strong advanced guard is a temptation to an engagement not previously contemplated. *The Germans in 1870.*

An advanced guard is divided into 3 portions, viz :— *Division into three portions.*

i. The advanced party, always preceded by 1 or 2 files under a N. C. officer. Its duty is to explore with cavalry patrols the roads in every direction, and in an open country to spread out like a fan. *Advanced party.*

ii. The support, about twice the strength of the advanced party. These two together are termed the vanguard, under the orders of the officer commanding the support, who is responsible for the proper road being followed; and by means of scouts he must ascertain that the country on each side is clear of the enemy. He keeps up communication with the advanced party by signals and connecting files, and verifies reports before transmission. *Support. Duties of officer commanding the vanguard.*

iii. The reserve, about $\frac{1}{3}$ to $\frac{1}{2}$ the strength of the whole advanced guard. *Reserve.*

An advanced guard ought to be provided with axes and intrenching tools. * *Party with axes and intrenching tools.*

If available, signallers should accompany, and a party, when possible, proceed with any important patrol, so as to give immediate information, and spare the men fatigue. The leading files must be carefully practised in, and made to repeat, the recognised code, as follows :— * *Signallers.*

By day: "Advance."—Wave of the hand or sword. * *Code of signals.*

By day: "Reinforce."—Bonnet waved, or held above the head.

"Withdraw."—Rifle reversed. (*Vide* Fig. 42.)

"No enemy in sight."—Rifle or sword held vertically, with bonnet on muzzle or point.

"Small bodies of the enemy seen."—Rifle or sword held horizontally steadily.

"Strong bodies of the enemy seen."—Rifle or sword horizontal, raised and lowered.

"Halt all in rear."—Hand held up high over the shoulder.

By night, or in foggy weather, or in thick woods, whistle sounds should be used. A prolonged whistle sound calls "attention;" and to men in extended order it means "cease firing."

On outpost duty, the cry of some bird or animal or other signals, may be pre-arranged.

Duties of connecting files. * Connecting files between the different portions have the important duties of transmitting signals from the front, and of showing the way to the troops in rear, by dropping a man at cross roads or other doubtful points.

Principle of formation. * One general principle applies to the formation of every advanced guard. Small parties of a few men are in front, supported immediately by others of increasing strength in rear.

The point. * The leading party is called the "point." Thus a constantly increasing resistance is afforded, the several parties marching at distances which are consistent

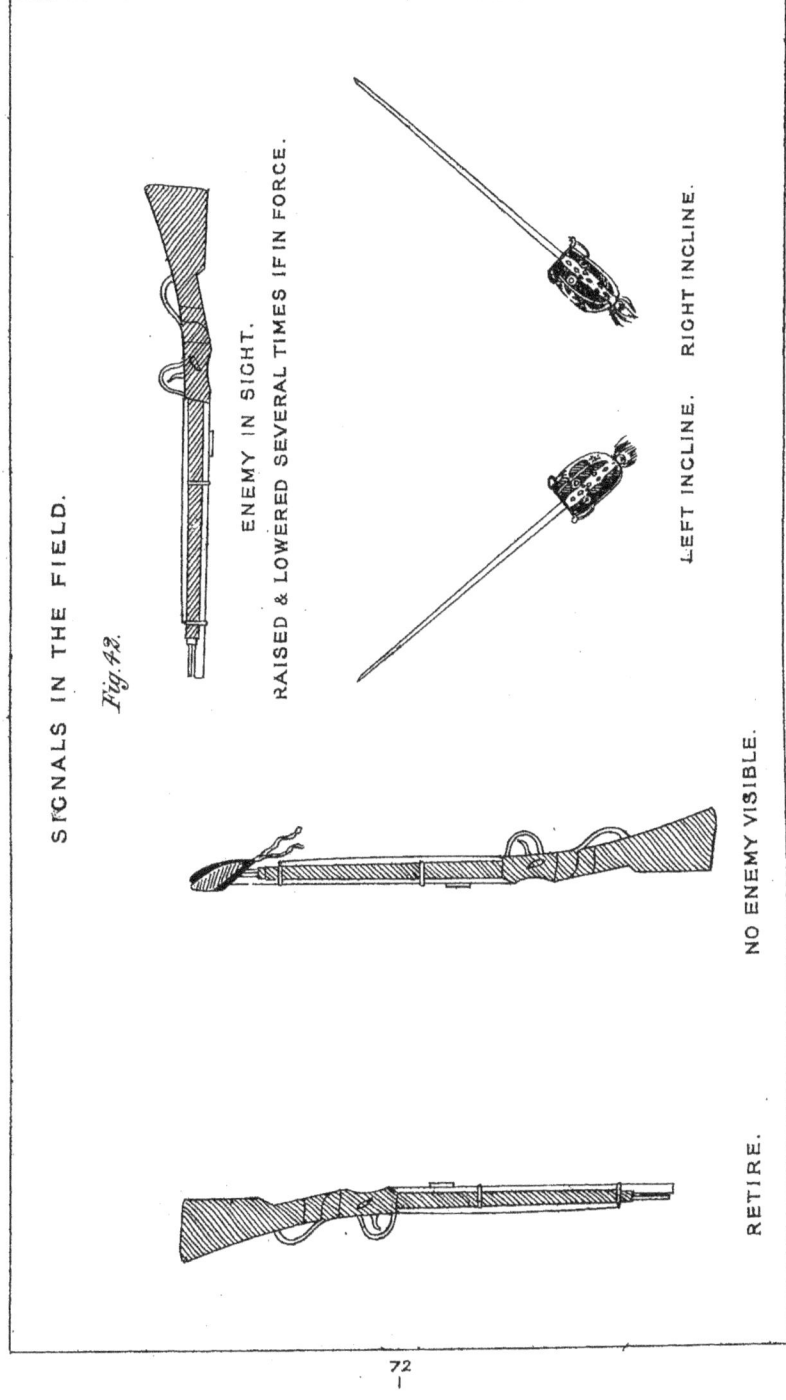

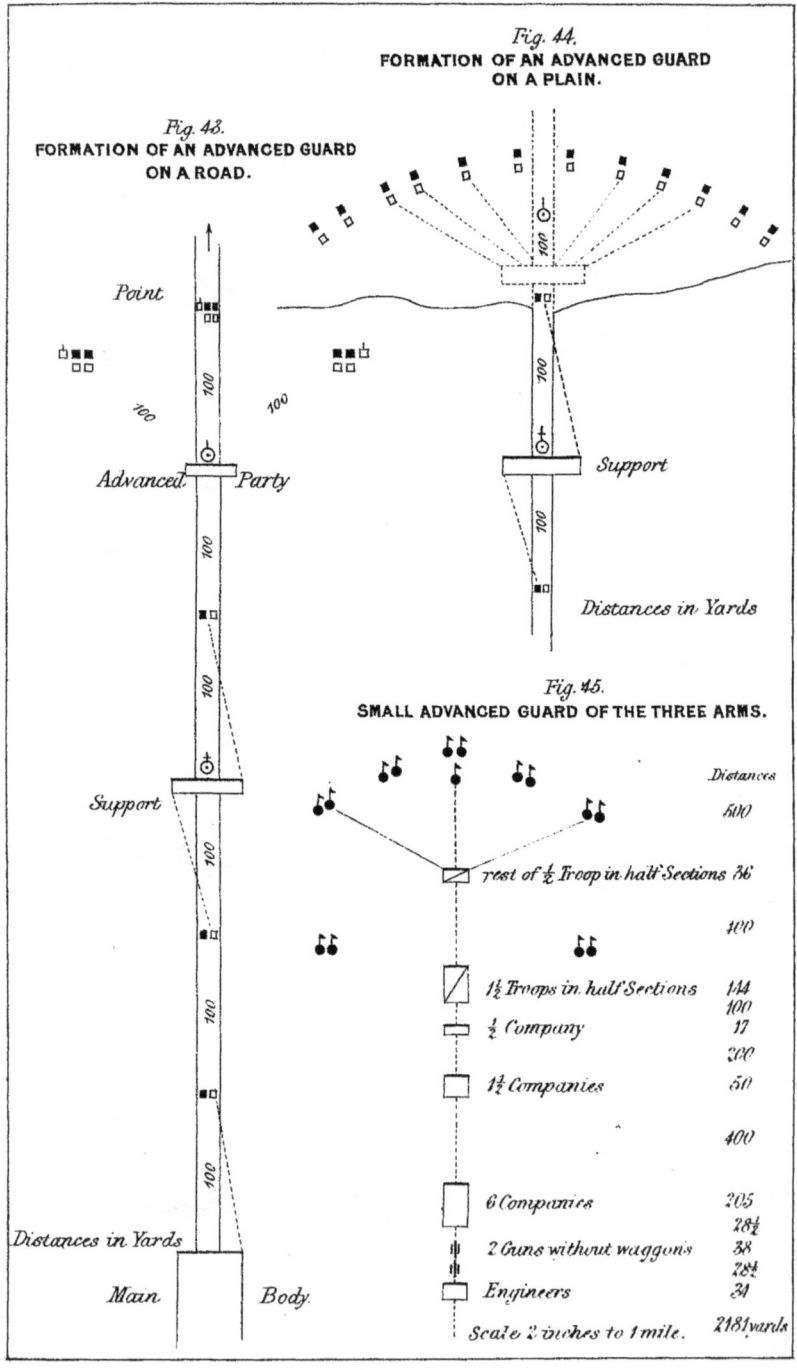

with connection and timely junction, and yet so far disseminated as to delay an enemy until the main body has time to form up for attack or defence.

The simplest case is that given in the *Field Exercises*, of the advanced guard of a battalion—usually a company—on a road. It is divided into two parts; one half-company under its guide detaches parties of 2 files 100 yards to the front, and 100 yards to the right and left front, each under a N. C. officer or selected soldier. If the company is weak, these parties may consist of only one file each. Any flanking party not required should join its half-company. The remaining half-company follows 200 yards in rear as support, and 300 yards in front of the battalion. It sends on a connecting file 100 yards, and drops another at the same distance. The leading company of the column also sends out one (*vide* Fig. 43). <small>* Advanced guard of a battalion on a road.</small>

When this advanced guard debouches from a road on a plain, the leading half-company is formed in extended order. The point halts, the flankers move up abreast, and lie down at the 100 yards' interval; the rest of the advanced party extend from the centre, and complete the line. Intervals are corrected from the centre on the march. The support follows, and the connecting file still communicates (*vide* Fig. 44). <small>* On a plain.</small>

When an advanced guard is composed of the three arms, cavalry will proceed to reconnoitre, as the fatigue and loss of time would be great if infantry performed this duty. In case the cavalry are checked, and to overcome obstacles, infantry and engineers will form part of the support. <small>* Reason for employing cavalry. Support of infantry and engineers.</small>

Of the main body of the advanced guard, a small part of the infantry will lead, followed by the guns, then the rest of the infantry and cavalry.

When guns may accompany and the minimum number.
When the road is hilly, and the troops must move with a narrow front, 2 guns without waggons may accompany the support, in order to contend with the enemy's artillery. This number of guns is the least that should be in position at the beginning of an action. By these means delay is avoided in sending back to the reserve for them; and no risk is run, as the enemy will not be able to extend his front more than the support.

Advanced guards of the three arms.
Fig. 45 shows an advanced guard composed of a battalion of infantry, a squadron of cavalry, and 2 guns.

Fig. 46 is one formed of an infantry brigade, a regiment of cavalry, and a battery of artillery.

Advanced guard of a cavalry regiment.
The strength of the advanced guard of a cavalry regiment would be about 2 officers and 60 N. C. officers and men, disposed in the open somewhat as indicated in Fig. 47; on a confined road 2 men precede as scouts.

Duties of the officer commanding the support.
The officer commanding the support, in the absence of special orders, acts according to judgment, providing individual supports for the separate advanced groups, or keeping the whole together on the main road.

The reserve and the main body each sends out a group to either flank. Connecting files—single or double—maintain communications between the several portions.

Distances between the fractions.
The distance is about 500 yards between the advanced groups, support, reserve, and main body, or,

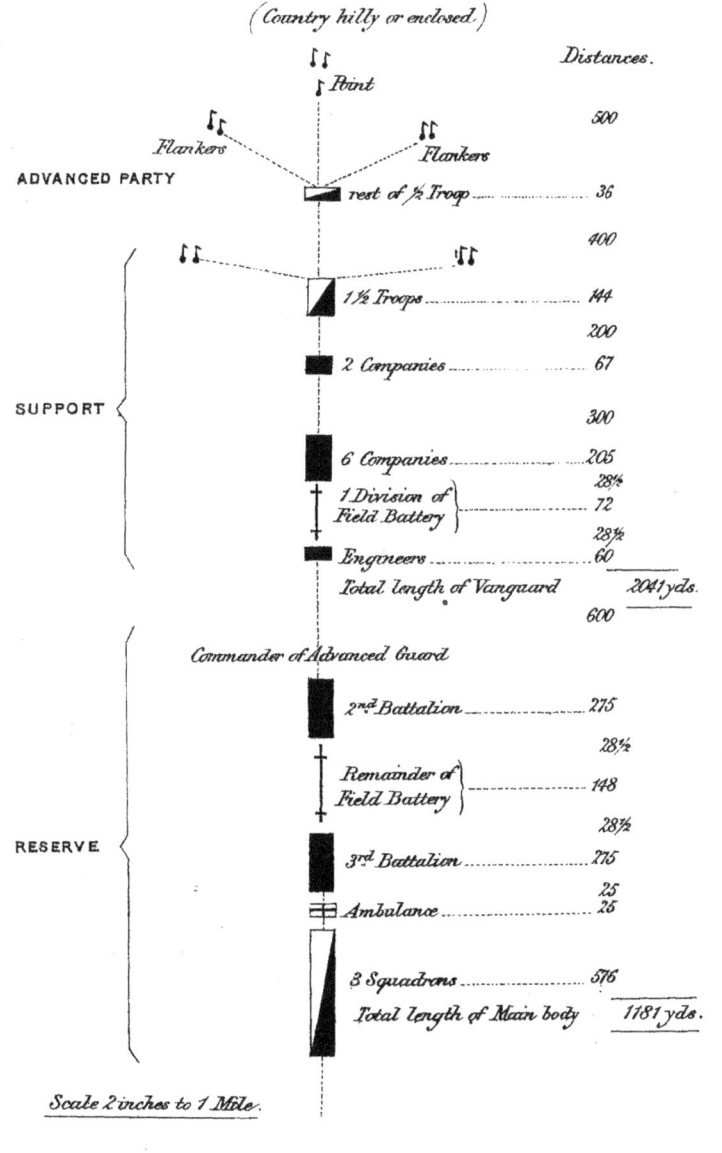

Fig 46. ADVANCED GUARD OF AN INFANTRY BRIGADE WITH A PROPORTION OF THE OTHER ARMS. (Country hilly or enclosed.)

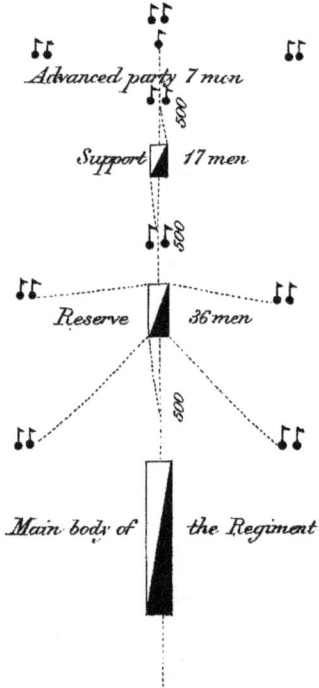

Fig 47.

ADVANCED GUARD OF A CAVALRY REGIMENT.

Scale 2 inches to 1 Mile.

roughly, a mile from the point to the head of the column.

Speaking generally, the head of an advanced guard should be so far in front, that the column, on first receiving information of the enemy, may have time to form for action before the enemy can reach it. The distance, therefore, depends upon the time the rear will take to come into line of battle. Still, it must not be so far to the front as to risk being cut off. _{*On what the distance of an advanced guard in front of a force depends.}

Houses, inclosures, copses, and other objects, capable of concealing an enemy on either side of the line of march, must be examined by detached files, and by patrols, when the object is too distant for examination by the flank files; *e.g.* a subaltern's party of 30 men, a sergeant and 12, a corporal and 6. Each of these parties should have a small advanced guard, and must never commit itself in action if avoidable. Such a patrol will move with a point, rear guard, and flankers on the most exposed flank. _{*Duties of flank files and patrols.}

On coming to a defile, a small party of cavalry should generally lead, to gain the first intelligence; then part of the infantry follows, to explore and secure the flanks. The same rule applies in case of a road bounded by strong walls or fences, a hollow way, a bridge in a deep valley, a wood, or a village street. Either of these might be fatal to cavalry or guns attacked therein by rifle fire, and unaided by infantry. _{*Manner of entering a defile.}

The flanking parties must occupy the heights on both sides before entering. If this is impossible, or there is no time, the N. C. officer at the head sends _{*Flanking parties.}

on a single man at the double, who keeps a sharp look out, and the others of the party follow in succession, keeping each other in view. The parties on the high ground on each side march ahead of the party in the centre throughout the length of the defile, and then gradually fall back to their original position.

Patrol entering a village. * A small patrol should not enter a large village, but reconnoitre it from the best position. The enemy must be felt for cautiously. The advanced party halts on the road, while strong flanking parties move round the outskirts to threaten the rear. The N. C. officer's party at the head advances along the road in single file at distances of a few yards, followed by the rest of the advanced party, and then the support, as soon as it is known that the place is not occupied.

If the village is small, the patrol may enter cautiously, having first seized an inhabitant for information. Two men should enter first, and passing down the road, keep their eyes fixed on the opposite houses. A third man remains at the entrance and keeps them in sight.

At night a patrol should hide close by to listen, and try to look in at the windows.

Examining a farm, &c. * A house or farm should be reconnoitred first from a distance, one man always remaining concealed close by for observation, while the others endeavour to enter from the front and rear simultaneously.

Examining a wood. * The same precautions are necessary to examine a wood or other inclosure. If the wood is of any extent, the borders only can be explored by a small patrol. One man moves first along the outer edge, while the others search it strip by strip.

A fixed rule is to turn the flanks, or threaten the rear of any place capable of concealing an enemy before feeling in front. Therefore, woods, ravines, and morasses must be searched before being passed, lest an enemy lying *perdu* succeed in cutting off the advanced guard. By threatening the rear before approaching the front, the enemy will usually be discovered, and dislodged without loss. * Always turn the flanks and threaten the rear.

So, before ascending a hill, the flank files on both sides go round the base, then a file from the point ascends cautiously, creeping when very near the top, so as not to show himself on the summit, but observing from behind the brow. He signals to the rear whether the enemy is in sight or not. * Ascending a hill.

The advanced party and patrols should ascend all heights, church towers, steeples, and high buildings, so as to obtain the best view, while keeping themselves concealed. * Best view to be obtained.

On coming to a bridge, they ought to examine it well to see if it has been tampered with, and then cross it as a defile. * Examining a bridge.

To explore a branch road, one man, followed by another, goes quickly to the first turn in the general direction, and comes back if nothing is seen. But if they discover anything suspicious, one of them should remain to watch, and the other runs back to stop the patrol on the main road. * Exploring a branch road.

The same principles apply to a cavalry patrol, but it can advance further, and explore a wider front than an infantry patrol. The scouts work in pairs. Usually half the strength of a patrol is in support, and the commander must keep the detached files Cavalry patrol.

Scouts and support. Touch of the enemy to be maintained. in hand by word or signal. A message should be sent back the instant the enemy is discovered; but the scouts should remain close to him for further information.

Although reconnoitring is a special subject in itself, the following concise instructions are applicable to patrols.

Instructions to patrols. * A patrol leader should have a map, and if going far, a guide. He should be able to answer these three questions:—

(1.) How far and in what direction am I to reconnoitre?

(2.) Where am I to look for the enemy?

(3.) What am I to do when I meet him?

Manner of making reports. * All reports should be in writing, showing the name of the sender, the hour, and place of despatch.

On the enemy's position. * Regarding the enemy's position, a patrol ought to note the strength and composition of his piquets, where they are posted, and if intrenched; whether the main body is close or at a distance; if the approaches are open or barricaded; if the outposts are vigilant, watching all points of passage; if guns enfilade special points; the uniforms of any of the enemy.

Avoid hasty conclusions. * Hasty conclusions are to be avoided, *e.g.* mistaking country waggons for artillery.

Roads. * As to roads, the chief points are their nature, width, gradients, and fences; if fit for all arms; together with full information on bridges, defiles, &c.

Railways. * Exploring a railway, notice the gauge, stations, platforms, sidings, arrangements for watering, tunnels, cuttings, embankments, bridges, rolling stock, &c.

In a wood, notice if it is open, as with pine or * Woods.
beech trees; or thick with brushwood; its extent;
if traversed by roads or paths, and where they
intersect; and if there are villages on the borders.

In the case of a river, note the width; strength and * Rivers.
direction of the current; the depth, whether constant
or variable; tidal or not; the nature of the banks
and country on each side; particulars of bridges
with means of repair, ferries, and fords; positions
commanding them. A canal is similar to a * Canals.
river.

In a town or village the chief points are:—its * Town or village.
situation, *e.g.* on level ground, on a height, in a
valley, on a river bank; the prominent features of
the surrounding country; if there is commanding
ground near; the nature of approaches; if the houses
are of wood, brick, or stone; of one or more stories;
thatched, tiled, or slated; close together, or scattered;
the principal buildings, as a church, town hall, manufactory; accommodation for men and horses; supplies
of provisions, corn, and forage.

There are no absolute rules for the commander of Duties of the commander of an advanced guard.
an advanced guard, who must greatly depend on his
own intelligence, and allow nothing to escape his
observation. He marches with the reserve. His
paramount duty is to prevent the main body being
attacked or its march interrupted. For this reason,
he should attack and disperse small bodies, bearing
in mind a maxim of Frederick the Great, "never Frederick the Great's maxim.
to haggle with the enemy's light troops." But he
must not engage seriously, if this is not in accordance with the plan of the general commanding.

Should he be opposed in force, he must select the most favourable ground for resistance.

<small>On first sighting the enemy.</small>
On first sighting the enemy, he should send back intelligence; and if the enemy appear weak, in the absence of special orders, attack at once, to prevent delay.

If the general purposes fighting on some position in rear, the advanced guard will retire slowly to it; but if he intends or is obliged to fight on the ground occupied by the advanced guard, the latter must hold on till reinforced, calling up every available man and gun to resist to the utmost.

<small>Advanced guard must protect main body at all risks.</small>
In extreme cases the advanced guard must sacrifice itself if necessary, rather than allow the enemy to push it back and open artillery fire on the main body unprepared.

Whether reinforcements are called up, or the advanced party is ordered to fall back on the support, will depend much on the ground, the enemy's strength, and the distance from the main body; the object being to give the column time to prepare for action without unnecessarily exposing the men.

<small>Should not pursue, but maintain the touch.</small>
Whenever the enemy is dislodged from any post or place, the advanced guard must be properly re-formed. It should not pursue, but simply maintain the touch of the enemy.

Flanking Parties.

<small>When necessary.</small>
* When a force on the march is exposed to attack on a flank, it must be protected by parties detached to that flank, moving parallel with the column, but when halted they front towards the enemy.

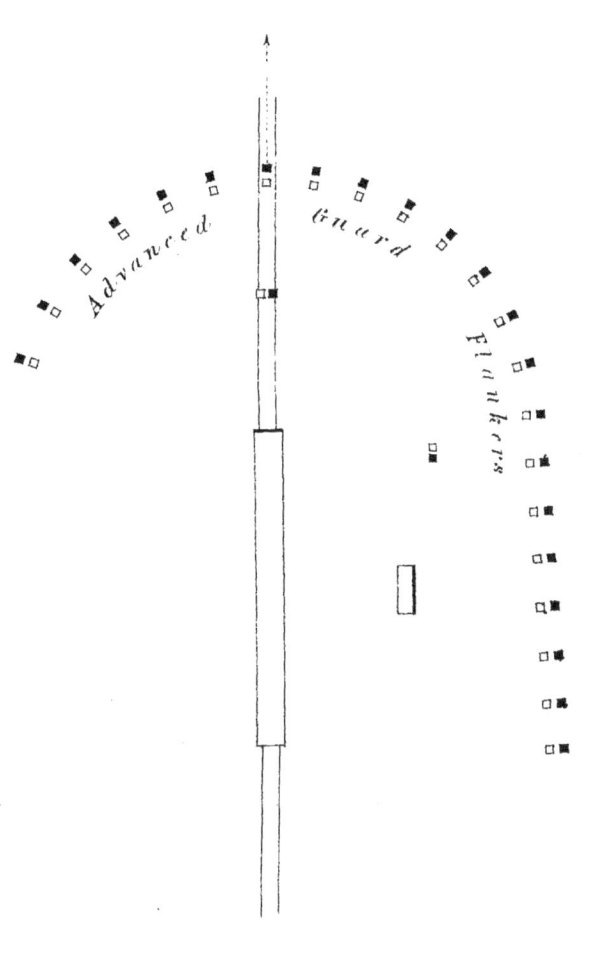

Fig. 48.

A BATTALION ON A ROAD RIGHT FLANK THREATENED.

A simple example is that of a battalion advancing on a road with one flank threatened. It sends out a company to cover that flank; half being extended at wide intervals, and connecting with the advanced guard, and half acting as support between the flankers and the column (*vide* Fig. 48).

* Protecting the flank of a battalion.

A Rear Guard

Is to be considered under two conditions: (*a*) to an advancing force, (*b*) covering a force retreating before the enemy.

The first is simplest, its duties being chiefly to act as police, protect baggage from plunderers, watch the waggon drivers, collect stragglers, and guard the rear of the column.

* To an advancing force.

But the second is very different. A rear guard covering a retreating army should be composed of the freshest troops, and formed immediately after the armies separate after an engagement. Its duties are to check and annoy the enemy, causing him to deploy, and so gain time for the main body to fall back in good order unmolested, or to re-form if in disorder, so that retreat may not be converted into a rout. This object should be effected, if possible, without fighting.

* Covering a retreating army.

* Duties.

It should be about $\frac{1}{8}$ to $\frac{1}{4}$ the whole force, a miniature army, composed of the three arms, much the same as an advanced guard in a forward march.

Proportion to the army.

Its distance from the main body depends upon the ground, and if the enemy is pursuing. It should be in constant communication with the column, to which it is kept closer than is usual for an advanced

Distance from the army.

G

guard. It commences its march as soon as all the waggons and baggage have moved off.

Formation.* Its formation is that of an advanced guard turned to the rear, and the proportions are reversed. The actual rear should be a line of skirmishers; and a support and a reserve marches between them and the column.

The rear guard of a battalion.* In the simple case of a company forming the rear guard to a battalion, it should first be turned about; and when halted its front is to the rear. The different portions should then be told off, or dropped in succession.

To be accompanied by a party with axes, tools, &c. A rear guard ought to be accompanied by a party carrying axes, intrenching tools, and explosives, to enable defiles to be blocked, bridges broken, roads cut up, or the enemy otherwise delayed.

If the retreat is on several parallel roads, each column must have its own rear guard.

The command of a rear guard a post of honour. Marshal Ney. The commander is chosen for skill and bravery. In a retreat he has the post of honour, and very important duties. Marshal Ney was renowned as a rear guard commander.

Duties of the commander. The men must be kept constantly on the alert; and when attacked, must defend obstinately every hedgerow, copse, or defile. The commander should occupy naturally strong positions across the line of retreat, and make a show of strength, to make the enemy believe his force to be greater than it really is. He should be kept informed of such positions as are suitable for defence, and of any obstacles the main body may meet with on the march, in order to be prepared to hold the enemy in check during the delay caused in passing such obstacle. Flanking parties must be instructed to be specially vigilant, so as to anticipate the enemy stealing round, and obstructing the main body, e.g. by a neighbouring ford, while the rear guard holds a bridge across the stream. *Flanking parties to be vigilant.*

As a rule the object is not to fight, but as soon as the enemy has completed his dispositions to attack, or when the flanks are threatened, the rear guard should quickly withdraw to the next selected position.

A rear guard retires under cover of an extended line. Artillery should open fire at longest effective ranges to delay the enemy. Horse artillery should be employed in preference to field artillery; and it is well for some of the guns to remain in action to the last moment. *How a rear guard should be withdrawn.*

The retreat is best carried out by alternate portions falling back from one position to the next. But touch with the enemy must never be lost, lest the pursuing advanced guard, moving rapidly by another road, overtake the rear guard, and cut it off.

The troops must fall back deliberately, and not allow themselves to be thrown into confusion.

The time for withdrawing from a rear guard action requires judgment. Night may be well chosen, and camp fires left burning to deceive the enemy.

But, on the other hand, a rear guard may be required to hold the ground at any cost for the sake of the army it is protecting.

CHAPTER VI.

OUTPOSTS.

<small>Meaning of outposts.</small>

<small>Result of their neglect by the French in 1870.</small>

* INSTRUCTIONS for outposts, advanced, and rear guards are comprised in Part VI. *Field Exercises for Infantry.* Every soldier should be fully acquainted with this subject.

Outposts are the "eyes and ears of an army," and just as a sentry over the guard-room door keeps watch while the guard sleeps, so an army can rest securely in camp or bivouac while the outposts are vigilant.

They have been aptly compared to the antennæ of an insect; and in the *Soldier's Pocket Book* the general form of the system is likened to a man's hand with the fingers well opened, the nails being the line of piquets, the middle joints the supports, the knuckles the reserve, and the wrist representing the army to be covered (*vide* Plate IV.).

* One cause of the disasters suffered by the French in their campaign against the Germans in 1870 was their having neglected the use of outposts. Thus, early on 4th August of that year, their troops were cooking, and otherwise employed, when suddenly a German battery opened fire upon them.

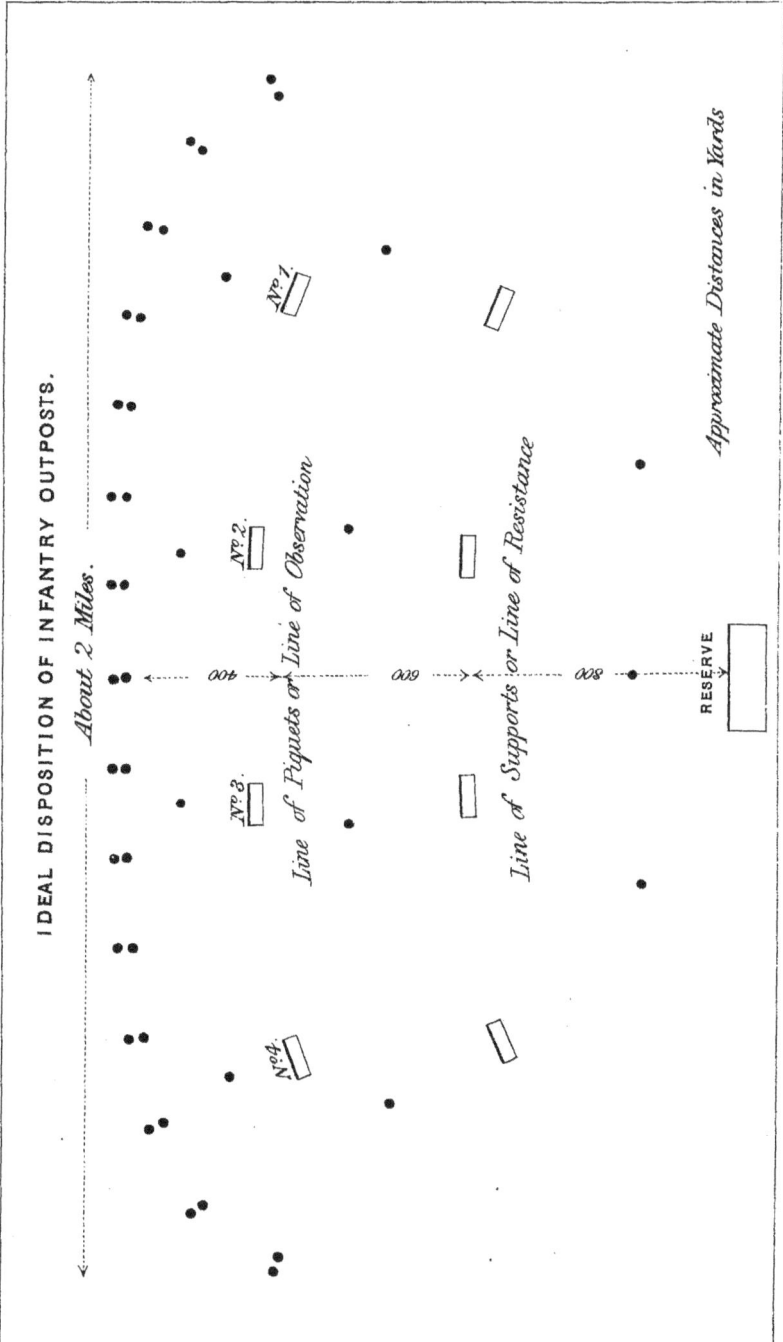

And a remarkable instance of a successful surprise and night march took place in 1811, when, in the Peninsula, General Hill, with 2 brigades of infantry, 2 regiments of cavalry, and some guns, surprised the French General Girard at Arroyo-des-Molinos. Hill marched at 2 A.M., and, though he had but a league to move, only arrived a little before 7 A.M. The night was wet and stormy. The French, having neither outposts nor patrols, were completely surprised. One brigade marched down the main street, with the pipes playing:—"Hey! Johnny Cope, are ye waking noo?" The other brigade marched to the rear of the village, and the whole French force was captured.

<small>Surprise at Arroyo-des-Molinos in 1811.</small>

Outposts are detachments thrown forward in front of a force that is halted, to protect it from surprise, reconnoitre the enemy, and hold him in check if he advances, thus allowing time for the force to prepare for action.

<small>* General duties and disposition.</small>

They are usually disposed in three lines, viz.—piquets, furnishing sentries and patrols; supports; reserve.

Ours is a combination of the cordon system and of patrols; the former being a chain of double sentries, intended to prevent individuals from passing through the lines; the latter for keeping out small bodies of the enemy, and giving timely warning of his movements.

<small>* Object of the cordon system;</small>

<small>* and of patrols.</small>

Speaking generally, outposts must vigilantly watch all approaches leading from the enemy, observe all his movements, and report them at once; if he advances, delay him to the utmost; and prevent the passage of spies and deserters.

Composed of the advanced guard or the freshest troops.	* As a rule, the troops composing the advanced guard, being some distance ahead of the main body, form the outposts; but if these have been engaged on the march, then the freshest troops should be detailed.
On what the composition of the outpost force depends;	* It is usual for the three arms to be employed; but the composition of the outpost force depends upon the nature of the country, and the proximity of the enemy. If he is distant, and the country open, cavalry are chiefly employed; when the enemy is near, and the country inclosed, the duty is best performed by infantry. But for the front to be
Usually of cavalry and infantry.	* thoroughly watched, both cavalry and infantry are needed: the former are pushed well forward, especially on the roads; while the infantry secure their
Signal stations.	* retreat, and afford them time to rally. Signal and look-out stations should be established, and a few
Mounted orderlies.	mounted orderlies distributed to communicate rapidly between the outposts and the main body.
Field Telegraph. Guns.	Sometimes the field telegraph may be employed. Guns, as a rule, remain with the reserve. They are useful to command bridges, defiles, or high roads.
	In all cases, the most suitable ground should be allotted to each arm.
Considerations determining the strength.	* Various considerations determine the strength of the outposts. For instance, whether they are intended merely to watch the enemy, and gradually fall back when attacked; or to oppose his advance on ground selected for defence; or whether the enemy can advance against the whole line, or any part, or against the flanks, or is restricted to a few approaches; what is the nature of the country, broken or wooded, defensible, and affording available

positions to observe the enemy; the position and enterprise of the enemy. A smaller force will suffice if there are defensible positions. But this must be increased to afford determined resistance to an enemy close at hand and vigilant.

Outposts ought not to exceed ⅛ the effective strength of the main body. But the number employed should be as small as possible, consistently with security, for the duty is most harassing. Proportion to the army.

The principle is for the piquets and supports to belong to the same battalion, and to consist of complete units, *e.g.* companies or half-companies under their own officers; the supports and piquets being of equal strength. * Principle on which the troops are detailed.

When a single battalion covers itself, the companies not required for piquets and supports form the reserve.

With a larger force the reserve is composed of separate troops (*vide* Plate V. page 105).

Outposts must be sufficiently in front of the main body to give the latter time to prepare for attack, yet not so far as to risk being cut off. In some cases they are pushed well forward to obtain a good view of the country beyond; but in an inclosed country they are nearer. * Distance in front of the main body.

The Germans, in their campaign against the French in 1870, employed from 300 to 800 men per mile on outpost duty. With us a strong battalion is considered sufficient to watch effectually a front of two miles, and this estimate tallies with the strength of the Germans at the Battle of Worth, viz., 400 men to a mile. * Number of men required per mile.

With a large force the general of division notifies

Allotment of ground to be watched. to brigadiers the ground to be occupied; and brigadiers point out to officers commanding battalions the sections of the line to be watched by each.

Component parts of the system. * The component parts of outposts are the chain of double sentries, piquets, supports when required, detached posts, patrols, and reserve, all under a special commander; and if the length of front be great, one such commander has charge of every three miles.

In the case of a battalion detailed for outpost duty, a staff officer points out to the commanding officer the extent of front to be occupied, probably indicating some conspicuous house, clump of trees, or other object to mark each flank.

Duties of the commanding officer. The commanding officer selects the companies to furnish the piquets, and makes each captain take down in writing the general line of outposts; the exact ground occupied by each piquet; what is known of the enemy's position and his movements; the directions most requisite to patrol; how to act when attacked; the extent to which a post should be fortified; how flags of truce, deserters, and others are to be received; the nature and frequency of reports; any changes that appear desirable at night, so that these positions may be examined by day.

By the aid of a map, or, if possible, riding over the ground before the arrival of the troops, he determines roughly the line to be occupied by the piquets and their sentries, and then chooses positions for the supports and reserve.

The "line of observation." * The line of piquets, following a range of hills, the edge of a wood, a river bank, or other natural feature, is the "line of observation," selected so that the sentries may have the best continuous view to the front, and the piquets may give timely warning of danger, affording the army security and rest.

On an average distance of 400 yards in rear of the piquets is the line of supports, determined by the most defensible ground available, where the first important stand is to be made. This is called the "line of resistance." * The "line of resistance."

The flanks, if possible, should rest on natural obstacles, e.g. a river or swamp. If not, they must be thrown back, and supported by detached parties, to anticipate a turning movement by the enemy; and the exposed flank should be frequently patrolled. * Flanks to be protected.

· The commanding officer further notices all roads and paths leading from the enemy, inspects the sentries and piquets when posted, altering their positions if necessary. He sifts all information, and reports it at once to the general commanding.

Each officer commanding a piquet, before moving off, inspects and takes nominal list of his men, examines the arms, ammunition, and rations, and sees that each N. C. officer has a pencil and paper or rough note-book. He marches his piquet to the ground pointed out with great caution, sending forward an advanced guard, with scouts and flankers as the ground may permit, taking every precaution for safety. On the way he should note all accidents of ground likely to be useful to make a stand in retreat, and the general features of the country. Arriving in rear of the intended line of observation, the piquet is halted, and a patrol or a small party of skirmishers is pushed on to explore the ground in * Duties of the officer commanding a piquet marching to the ground.

* On arrival.

front. Under cover of these he posts the first relief of double sentries, which, in advancing, open out by files, so as to cover approximately in extended order the whole ground allotted to the piquet to guard. On arriving at the line of observation, if nothing is seen of the enemy, patrols proceed still further, to examine doubtful ground within easy rifle range, and the officer places the sentries in the best situations for seeing without being seen, and communicating with the sentries of the piquets on his flanks. He should pass from flank to flank in so doing, along the line of sentries, correcting their positions, and, if possible, reducing the number of posts.

<small>Another way when the country is open.</small> * If the country be open, a piquet may advance to its ground with half the men extended, supported by the remainder, and on reaching the position, the officer can withdraw all men not required as sentries from the extended line.

Each piquet can watch from 800 to 1,000 yards, or roughly, half a mile of front.

<small>Outposts remain under arms and alert till all posted.</small> * All outposts must remain under arms and on the alert until the whole system is completed.

<small>What determines the strength of a piquet.</small> * The strength of a piquet depends much on the nature of the ground. The sentries will be fewer in an inclosed country, and more men will be employed patrolling. The number of double sentries will rarely exceed 4, and these require 3 reliefs. A single sentry is posted over the arms; and a connecting sentry—also single—is sometimes necessary, as when all the double sentries cannot see the piquet. These are in 3 reliefs. The rest of the men are detailed according to circumstances, *e.g.* to

furnish an examining party, or a detached party, and patrols. A piquet is about the strength of an ordinary company, or half that of a strong one; usually from 25 to 50 men, and rarely exceeds 60, for more than this number is unmanageable.

In considering the distribution of a piquet in an open country—strength 1 captain, 2 subalterns, 4 sergeants, and 60 rank and file; total, all ranks, 67 —required to furnish 2 double sentries by day, and 3 more by night, a connecting sentry, a small detached party, and 2 patrols, the men must be selected for each duty according to their capabilities, and the company will be divided somewhat as follows :—

* Example of the distribution of a piquet.

For 2 double sentries by day in 3 reliefs	12
,, 3 ,, ,, night ,, ,,	18
,, 1 single connecting sentry ,, ,,	3
,, 1 ,, sentry over arms ,, ,,	3
Detached party, 1 subaltern, 1 sergeant, and 9 men	11
Patrols, 2 sergeants and 7 men	9
Remainder for duty, 2 officers, 1 sergeant, and 8 men	11
Total	67

A couple of men, with a field-glass if available, should be posted on an eminence, in a high tree, or tower.

Sentries ought not to be too close to their piquet, lest the enemy attack them and the piquet at once, and the latter will have no time to form for resistance. The first posting is tentative, the object being to have the whole front watched, and communication from flank to flank rapidly established. Before deciding whether a man is posted to the best advantage or not, an officer should place himself exactly in the position of a sentry. Piquet sentries

* Sentries first posted experimentally.

* The best way to post them.

should have a wide view to the front and flanks, and command all approaches. No ground between two sentries should be unseen by both. Connection must be maintained with the sentries of piquets on the right and left. Their number may often be economised by taking advantage of features of ground. All should be concealed from view of the enemy.

Sentries to be economised.

An officer must satisfy himself that his sentries understand and can answer the following questions:—

Questions for sentries to know how to answer.

i. What is known of the enemy? and, indeed, the whole of the piquet cannot be too soon, or too well, informed on this point.

ii. Where are the neighbouring piquets, any detached or advanced parties, and the other sentries?

iii. Where do the roads in sight lead to?

iv. Do they thoroughly understand the code of signals fixed upon?

v. Are they sure as to the direction their attention is to be devoted to? This means the front of their posts; and that they should know this better a row of stones may be placed in the direction, or a pointed stick laid on two forked props. On a dark night a white rag should be tied round the sight of the rifle.

A sentry must further have pointed out to him the position of his piquet, and means of access to it, so that he may be able to rejoin it at night and in all weathers.

Connecting sentries.

If he cannot communicate with the piquet, a connecting sentry is placed so as to see both. Such sentry must repeat all signals. Sometimes a connecting sentry is requisite to maintain communication between the chain of sentries and an advanced or detached post beyond the line.

A piquet sentry has no fixed beat or patrol. His duty is properly performed if he sees well to the front and flanks, and can communicate to the flanks and rear by signal in the open, or by word of mouth in an inclosed country.

The front and rear rank men of each double sentry need not necessarily be together. The ground may be better observed in front sometimes by placing them a few paces apart, as when a zigzag path in a ravine, or a lower ridge of a hill, cannot be well seen from the crest. * The two men of a double sentry not necessarily together.

It must be explained clearly to a sentry what to do in falling back. He must not fire unless attacked.

If posted in a narrow way, or other place where he might be surprised, the post should be protected by an obstacle, and he should fix bayonets in thick weather, or on a dark night, but not by moonlight. * Sentry to be protected against surprise.

A sentry must take care not to be deceived by parties approaching his post, even though preconcerted signals have been arranged for the use of visiting patrols.

As far as possible, the same men should mount constantly on the same posts, and men for patrolling employed in the same direction as that in which they were first sent out. * Same men to mount same posts.

Sentries posted with valises on are not to take them off. Nor may they lie down except for concealment. * Sentries not to take off valises.

The best postion for sentries posted in a wood is just inside the outer edge. But if the wood is

Where sentries should be posted in a wood.

* extensive, it is preferable not to occupy it, the sentries being withdrawn a safe distance—say 600 to 800 yards—on the near side, under some made cover. In this case the wood must be constantly patrolled.

If it is imperative that the line pass through a wood, the sentries are best posted along a high road, a stream, or a ravine.

Usually relieved every 2 hours.

* Sentries are usually relieved every 2 hours. In severe weather they may be relieved every hour.

On being relieved they should be questioned as to what they have seen, and in answering they should not conceal even the most trifling occurrences.

Duty of sentry on the enemy's advance.

* When a sentry observes the enemy advancing, he must signal to the piquet. If it is plain that an attack is intended, he must fire at once, and use all means to alarm the piquet.

Position of sentry by day.

* By day the best positions for observation are on high ground, the sentries taking advantage of natural cover, or lying down just behind the crest of hills. As few changes as possible should be made at night, and these determined on during the day.

By night.

* At night sentries must be advanced down the slopes towards the enemy, in order to see him on the sky line without giving up the advantage of the high ground, while remaining concealed themselves. The shadow of a tree or wall is a great advantage at night.

Simple signals to be prearranged.

* Simple preconcerted signals should be arranged (*vide* the code under "Advanced Guards," Fig. 42, p. 72); all the piquet must learn them, especially sentries, connecting sentries, and patrols. It must be impressed on every one on outpost duty that information is first obtained by observation, and when so obtained it has to be transmitted.

Signalling by flags may often be made use of. A few mounted orderlies are invaluable; and generally information is more reliable and more quickly brought in by them.

On the approach of deserters, they must be ordered to lay down their arms, and be conducted unarmed to the piquet. * Deserters.

If one of the piquet deserts to the enemy, a report of the fact must be at once made to the commander of the outposts; and the parole and countersign should be at once changed. Precaution when one of the piquet deserts.

A flag of truce must be kept at a distance in front of the line of sentries, while the piquet is communicated with, so that the bearer can get no information or reconnoitre. If he brings a letter, a receipt should be given and the bearer forthwith ordered to depart. In other cases the officer commanding the piquet will detain the flag of truce until he reports to the commander of the outposts, or he will forward the party blindfold under escort to camp. * Flag of truce.

No one is allowed to pass the chain of sentries except at the post of the examining party. Any one who disregards a repeated challenge should be shot.

An examining party, or, as it is sometimes called, an examining guard, consists of 4 or 6 men under a N. C. officer, posted on a main road, running centrally through the position; or, in case of more than one such road, there may be a party on each. Their duty is to examine persons wishing to pass the chain of sentries; giving or withholding permission; or * Examining party.

reporting a particular case, awaiting instructions from the commander of the outposts.

If there is no examining party, the officer of the nearest piquet performs this duty.

It is to the examining party that piquet sentries should direct all persons approaching the line. On coming to the examining party, one man halts the individuals before they can overlook the line of piquets, while another sends word to the N. C. officer. They should be questioned when halted, or blindfolded before allowed to pass. Deserters or suspected spies are sent on at once without questioning to the commander of the outposts, or dealt with specially as may be ordered in each case.

Detached * post, when necessary. If a point is difficult of access from a piquet, or at some distance beyond the general line, and it is incumbent to occupy it, or when it is necessary to protect an exposed flank, a detached post is established under an officer, or a N. C. officer, with sufficient men to furnish the required sentries. The post should be relieved every 6 hours. The commander is responsible for the communication between his post and the piquet. His orders should be carefully explained to him and the men before starting, as also the object for which they are detached. If such a post is composed of infantry, the sentries should be posted close in front of the party, the reliefs remaining equipped and having their rifles close by them as they sit or lie on the ground. No fires are allowed. If cavalry are employed, their vedettes may be 300 or 400 yards in advance. Only one half should dismount at a time, unless the enemy is known to be at a distance.

Communication is kept up between cavalry detached posts by means of patrols.

When a river or swamp extends along part of the line sentries may be decreased, and observation maintained by patrols, thus giving rest to as many men as possible, consistently with safety.

An officer commanding a piquet, having completed his arrangements for the sentries, returns to his piquet, piles arms, and posts a bayonet sentry over the arms, and connecting sentries if necessary. These must keep their attention constantly fixed on the advanced sentries. He should send a report of preliminary dispositions to the commander of the outposts, and direct his second in command with reference to any intrenchment or temporary strengthening of the post which appears desirable, and as to what patrols are to be sent out. If his piquet furnish a detached post, he should now visit it, to assure himself that it is in the best possible position, connected with the main line, with reconnoitring patrols in front, and strengthened against a sudden rush upon the sentries.

* Further duties of the officer commanding a piquet.
* Sentry over arms.

Having written down his instructions, his general duties are to learn all he can of the surrounding country by means of a field-glass, and comparison with a map; to reconnoitre all paths and roads, noting their direction. But he should not go beyond reach of the piquet. If he has time and means, it is well to make a rough sketch of the post. He should examine all obstacles and places likely to conceal an enemy, and find out where marshes and streams in his front can be crossed. Let everything be done quietly. No drum or bugle

H

sound except the "alert" is permitted. The men should not be allowed to stray from the piquet. All first arrangements are subject to modification by the commander of the outposts or by an officer commanding the piquet himself after careful inspection. There are no hard and fast rules: an arrangement applicable to one tract of country might be quite unsuited to another. But the principle to recollect is to watch the country thoroughly in front, and economise sentries by taking advantage of ground. And an officer must make up his mind how to act whatever may happen.

<small>No hard and fast rules on outpost duty; but certain fixed principles.</small>

An officer in command of a piquet should be provided with a field-glass, a compass, a map, a watch, a note-book with pencil, and a few simple sketching materials.

<small>Piquets numbered.</small> * Piquets are numbered from right to left, so that an officer or N. C. officer, to whom a report is made, can identify the exact place in the line of outposts whence it comes. The men should be made acquainted with the number of their piquet.

<small>Piquet report.</small> * A piquet report should state the number of the piquet it comes from; the hour of despatch; how the information was obtained; if considered true; and all particulars of the enemy's troops, if seen.

<small>Piquets generally may not light fires.</small> * As a rule, piquets are not allowed to light fires, for if they do not take cooked rations with them, these are sent them from the rear. If a fire is

<small>Precautions when a fire is allowed.</small> * permitted, it must be carefully screened behind a wall, bank, or other cover, so that neither light nor smoke is visible from the front. Wet sods or earth should be handy, to extinguish the fire at a moment's warning.

An alarm post should always be fixed a short distance in rear of the fire, affording concealment, so that the enemy will be exposed in advancing. *Alarm post.*

The men of each relief should pile arms together; and the reliefs be kept separate from the patrols and from each other, so as not to disturb the rest when going out. All the men, or part only at a time, are usually allowed to take off their valises. Every man should keep his own close beside him. *Reliefs kept separate.*

If an attack is expected, part of the piquet must be kept under arms. The whole must invariably be under arms before daylight. *When all piquets should be under arms.*

With regard to its sentries, vedettes, or detached parties, a piquet is an anchor or rallying point. It should be posted centrally about 400 yards in rear of the sentries it furnishes. But it cannot always be so placed, because it must command all approaches, and be under cover if possible. Its best position, therefore, is near a road, concealed, but ready to move in any direction, communicating with its sentries, the piquets on its flanks, and the supports. Above all, it must have a good line of retreat. *Position of piquet as to its sentries.*

The general disposition of piquets depends on the configuration of the ground, the atmosphere, the vigilance and proximity of the enemy. Roughly, piquets may be from 600 to 800 yards apart; but in an open country and in clear weather they may be further.

Another consideration is, they must be sufficiently near to afford mutual aid to one another in retreat by an efficient flanking fire. They should not be too close in front of the supports, for fear of demoralising the latter if driven in, nor too far distant to *Piquets must afford mutual aid.*

H 2

receive timely assistance if hard pressed. By a piquet being concealed near a road likely to be used by the enemy in his advance, the enemy must attack in order to discover it.

Should not occupy buildings without orders. A piquet should not occupy a building or inclosed yard or garden, unless ordered.

Posts to be strengthened. An officer commanding a piquet or support must strengthen his post when practicable by a breastwork or abattis. Some such means should be extemporised

At a bridge or ford. to protect the men when defending a ford or a bridge, the piquet being posted on the near side, while the far side, if not too distant, is watched by sentries or

At a small village or defile. patrols. The same is true if the piquet is at a small village or defile. If a piquet is posted in a defile, it must be protected from a sudden rush by obstacles.

Bridge not to be broken, nor main road blocked. A bridge should not be broken, nor a main road blocked with material difficult to remove, as the army may have to use such bridge or road.

Flank piquet unprotected to be supported. A flank piquet, unprotected by a natural obstacle, should be rather withdrawn, and supported by a detached party.

A piquet may occupy a wood if thinly planted, and there is a clear view of the front from the far edge, along which the sentries are placed. The piquet should be near the sentries, posted centrally.

Piquet in a wood. If this is not possible, it is best to withdraw both piquet and sentries to the near side to guard against surprise, and explore the wood frequently by patrols. Instances, however, may occur when it is necessary to run the line through a thickly wooded district, and then the piquets are best placed along a road, stream, line of hills, &c., traversing the wood laterally. Many and small piquets are here preferable to a few strong ones; and the supports must be brought much closer.

A point regarding roads deserves notice, viz.:— When roads from the front converge, they should unite in advance of the piquets, and roads passing round the flank to the rear should be carefully watched. *Best point of convergence of roads*

The rule is for supports to be furnished by the same corps as the piquets; and the supports together should be of equal strength to the piquets. Thus 4 companies on piquet will have 4 in support. The number usually depends on the number of main approaches to be watched. There need not be a support to every piquet, but often there is one for two or three piquets, placed centrally in rear of them, on or near a main road. *Supports equal in strength to piquets. Not necessarily one to each piquet.*

Their distance in rear of the piquets varies from 400 to 600 yards. The principle is to have them near enough to be useful, but not so near as to be involved in the retreat if the piquets are driven in. But the supports might in exceptional cases be much nearer, or even on the very same ground as the piquets, when this affords the best defensible position. A suitable position for a support is a village or defile on the line of retreat. *Principle deciding their distance in rear.*

By means of patrols and connecting sentries a support must communicate with the supports on its right and left, the piquets in front, and the reserve; a bayonet sentry, mounted over the arms of a support, keeps constant watch on the piquets in front.

The strictness of routine insisted on with piquets is somewhat relaxed with supports. But they must *Strictness of routine relaxed with supports* be ready to march at a moment's notice, by day or night, to any required point, or to remain on the defensive. They can generally light fires and cook. In this case they prepare dinners for the piquets. But if no fires are allowed, and no cooked rations are carried, the food must be sent up from the reserve. When practicable, hot food should be provided for troops on outpost duty.

<small>Changes for night to be arranged by day.</small>

* Changes in dispositions for night duty should be determined during the day. All ranks must be acquainted with the ground, so that sentries may watch well, and be regularly relieved, and patrols may move with confidence, and no unnecessary fatigue.

<small>Dispositions of piquets by night.</small>

* At night the piquets occupy all obligatory points of passage along which the enemy may attack. In an inclosed country instead of furnishing a continuous chain, the double sentries are posted to watch all roads, paths, and avenues leading from the front, with reliefs close by to signal the enemy's advance; as in attacking he will confine himself to the approaches. But the intervening ground must nevertheless be constantly patrolled by small parties, of a N. C. officer and a couple of men.

If the country is open, there must be more double sentries, all pushed down to the low ground, and then patrolling will not be so much needed.

In some cases it may be desirable to contract the front at night, and bring the supports closer. The original positions are resumed at daylight.

<small>Reason for outposts being relieved at daybreak.</small>

* Daybreak being the hour when attacks are generally made, the utmost vigilance must be exercised at that time. For this reason outposts are relieved at daybreak, so that both the old and new piquets may be available to resist.

On being relieved, the officer commanding the old piquet accompanies the relieving officer round the chain of sentries, points out the locality, and gives him all information in his power. If everything is quiet in front, and the weather clear, the old piquet falls back to the support, and then returns to camp. But if on the way firing is heard, it must instantly turn back, and assist the new piquet. *Duties of officer commanding old piquet on relief.*

An officer commanding a piquet must make up his mind definitely what to do if attacked, and choose a line of retreat that will not mask the fire of the supports any longer than necessary. On being first attacked he will reinforce his line of sentries with part of or all his men, and send word or signal to the adjoining piquets and to the supports. The sentries hold their ground till their flanks are turned, then fall back, step by step, in extended order, taking care to keep clear of the other piquets and of the supports, so that the enemy may be met by a direct and flanking fire. The piquets retire slowly, in échelon, or alternately, on the supports, who by this time should be judiciously extended on the strongest available ground. *Conduct of piquet when attacked.*

If the enemy is repulsed or withdraws, the piquets advance and take up their old ground. They must not pursue, but a cavalry patrol should follow him up, see where he halts, and maintain the touch.

In the event of a night attack the sentries fall back on the piquets, who, being carefully posted beforehand, ought to be able to meet the enemy advantageously. *During a night attack.*

|Position of * reserve.| The reserve forms a general support to the whole outpost line of piquets and supports, and should be from $\frac{1}{8}$ to $\frac{1}{2}$ the strength of the troops employed, posted from 500 to 1,000 yards in rear of the supports, centrally and out of sight; not necessarily in one body, but it may be subdivided to hold important points, e.g. two defiles or bridges on different lines of retreat. The men may bivouac, cook, eat, smoke, and rest, but must be always ready to act at a moment's notice.

The officer commanding the outposts remains with the reserve. All reports should be sent in to him.

The reserve must be in constant communication with the outposts by signallers, mounted orderlies, and the field telegraph, which last usually conects it with head-quarters.

Composed of the three arms. The reserve is usually composed of the three arms. Artillery is useful when it can be posted within cannon range, but beyond rifle range, of any place where the enemy must advance in column before deploying.

Guns. Guns should be limbered up near a main road, but never in an inclosure. For the defence of a defile they should be unlimbered ready for action. They are sometimes posted with the supports and piquets, but if far advanced are liable to capture. They ought to command effectually any bridge or defile by which an enemy must advance, protected if necessary by a small body of infantry extended in front, or to one flank.

The general duty of the reserve is to assist the supports and piquets, or to occupy a previously selected position with a view to covering their retreat.

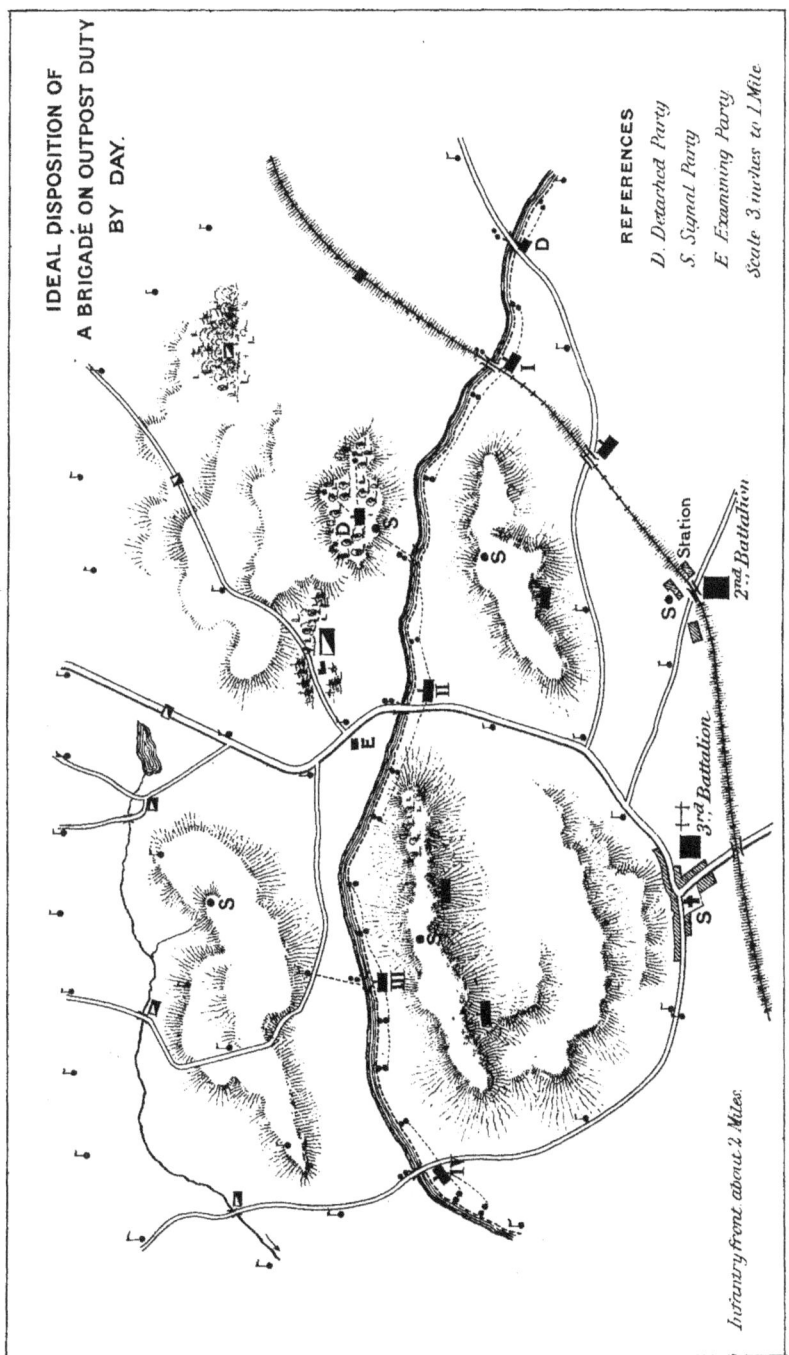

Plate V. gives some idea of the disposition of a brigade of the three arms on outpost duty by day.

The outpost force should be as far in advance of the army it has to cover as possible consistently with safety. This distance depends upon the position of the ground chosen for the "line of resistance." There is no fixed rule, but the object to be kept in view is the delay of the enemy beyond the effective fire of artillery, which may be taken roughly at 2,000 yards, so that the camp may not be shelled by the enemy.

Sentries on outpost duty pay no compliments. If a superior officer visits a piquet the officer in command of it reports to him, but the men take no notice unless addressed. On the approach of the commander of the outposts, the men stand to their arms only when so ordered. [* Sentries pay no compliments.]

Three kinds of patrols are to be noticed as employed on outpost duty: (1) visiting, (2) exploring or reconnoitring—these two are furnished by the piquets, (3) strong patrols, of a strength exceeding 12 men, sometimes sent out from the supports or reserve. Of these seriatim:— [* Three kinds of patrols.]

A visiting patrol goes round the sentries between reliefs to test their vigilance, ascertain if they are alert, if they have anything to report, or require assistance to examine doubtful objects, and to relieve cases of sickness. It consists of an officer, or more usually a N. C. officer or old soldier, and one or two men. Such patrol encourages the sentries and maintains communication between them and the piquet. Its usual route, after communicating with the nearest sentry of the piquet on one flank, is to [* Visiting patrol.]

proceed cautiously along the front to the nearest sentry of the piquet on the other flank, returning to its own piquet by the rear (*vide* Fig. 49).

Object of signals. * In order that the sentries may distinguish the patrol from foes a preconcerted signal should be arranged to be given, *e.g.* a cough, a whistle, the cry of a wild bird by night, or a handkerchief held in the hand by day. The sentries will not then be disturbed. The signal should be acknowledged by the sentries.

When patrolling is most required. * Such patrolling is the chief guarantee of safety, and is especially necessary in a close country, in bad weather, and at night. In foggy or thick weather sentries must be constantly visited, and all woods, ravines, and places where an enemy could collect must be examined, sometimes specially. Patrols may be less frequent in an open country, and in clear weather, or if the line of sentries can be overlooked from some point near at hand, and reports are often obtained from thence as to the state of things.

If the piquet is weak, reliefs must act as visiting patrols.

Exploring or reconnoitring patrols. * Exploring, or, as sometimes called, reconnoitring patrols, are best adapted for examining ground not visible to the sentries, and for giving timely notice of the enemy's advance. They are sent along roads for ¾ to 1 mile towards the enemy, and consist of the smallest number of men that will perform the duty; but they must be the smartest and most intelligent. Their duty is to reconnoitre, and not to fight.

Mode of proceeding. * Proceeding along roads they should listen from time to time for sounds of men moving through

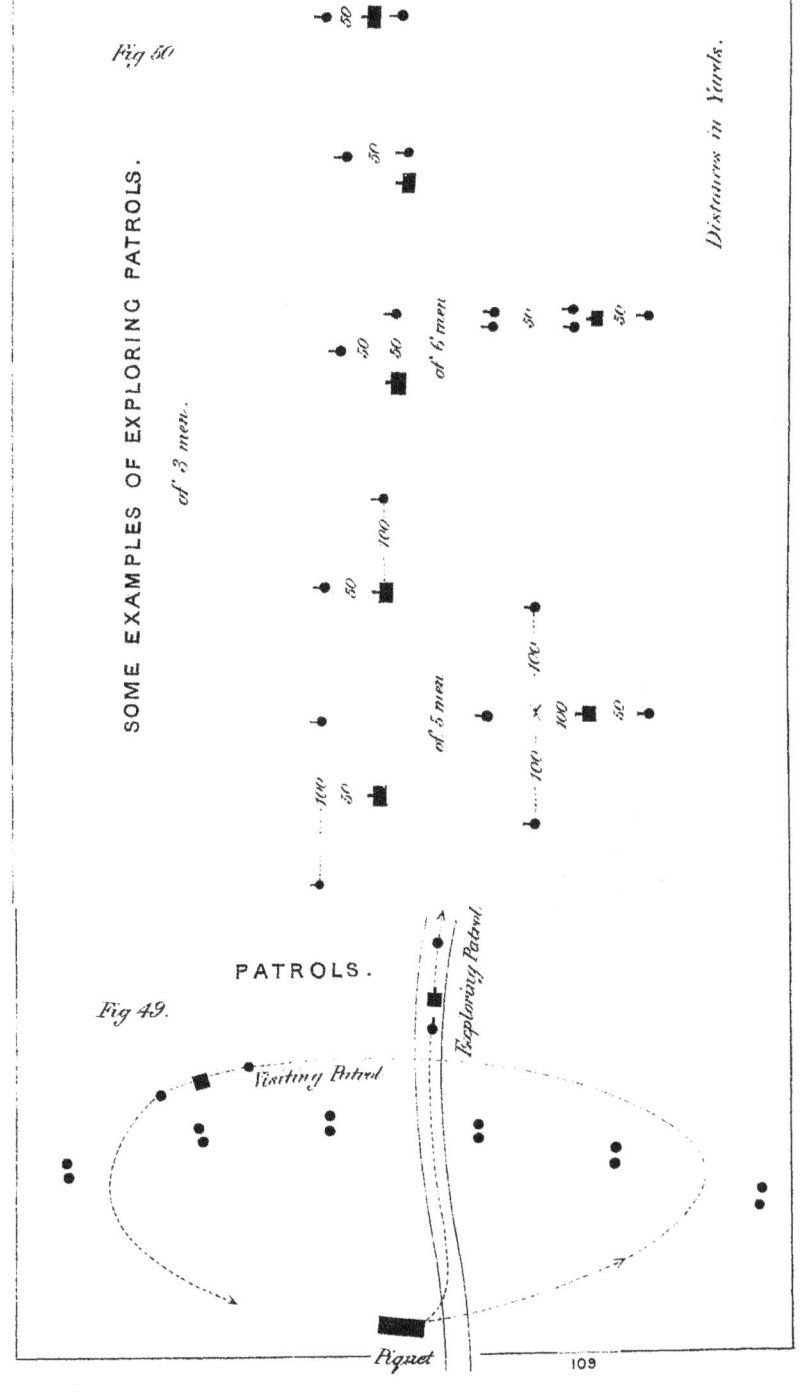

the fields or woods. On meeting the enemy, they should avoid firing, but if he advances they must fire to warn the piquet. They may not talk, smoke, or allow their arms to clatter. They ought to notice landmarks and other objects, and make marks with stones, or by breaking a branch, &c., so as to find the way back. If possible, it is best to return by a different way. Such a patrol moves like a small advanced guard. It should not be less than 3 men. There are many formations, according to ground. A point protects the front, and flankers the most exposed flank, at from 50 to 100 paces distance, so as to see best without showing themselves, supporting and assisting each other, yet not being so close as to risk all being cut off and captured (*vide* Fig. 50). One man moves behind another along the hedgerows, by the sides of roads, and on soft ground in preference to the centre. Their difficulty is to guard against surprise in a close country, and against being seen in the open.

When a small patrol reaches a hill, one man ascends the slope very cautiously and looks over the brow; another follows at a distance, and then the rest, so that all may not be cut off. But when possible the flanks of all objects should be turned before venturing beyond. For further instructions as to these patrols, *vide* " Advanced Guards." * Ascending a hill.

* Flanks to be turned.

Whatever information is obtained must be transmitted to the rear by means of the established signals, without attracting the enemy's attention, and to be of any value it must be early, ample, and accurate. * Transmission of intelligence.

A strong patrol resembles a reconnoitring patrol,

A strong patrol. * but is stronger, and not necessarily secret. It is sent out when necessary from the supports or reserve, and may be a troop or company. Its object is to press back the enemy's reconnoitring patrols, and by engaging a post to try to dislodge it, so as to ascertain what is behind, and what the enemy is doing. Although acting offensively, it ought not to advance beyond a mile. A mounted man should accompany, to bring back information.

A strong patrol is occasionally sent out before daybreak. Great caution is then needed so as not to come unawares upon the enemy forming for attack at daylight.

Reconnaissance in force. It may here be remarked that a reconnaissance in force, composed of the three arms, is sometimes pushed forward to find out the enemy's strength and position, and is often followed by an engagement.

Patrols to avoid needless firing and false alarms. * Patrols should avoid unnecessary firing, unless the enemy is advancing in force, and the troops in rear must be aroused. Every one on outpost duty should carefully avoid giving false alarms, which break the rest of the main body, and cause inattention to signals if really attacked. The most prudent course for a strong patrol is to retire steadily unperceived, after obtaining touch with the enemy, gaining the requisite information ; and, if so ordered, taking a few prisoners for further knowledge.

Cavalry piquet. As outposts are frequently composed of both infantry and cavalry, it is well to know that a cavalry piquet, consisting of from half a troop to a troop, will watch about a mile of country, furnishing, as a rule, not more than 3 pair of vedettes about 600 yards in front of it. Half only of a cavalry piquet

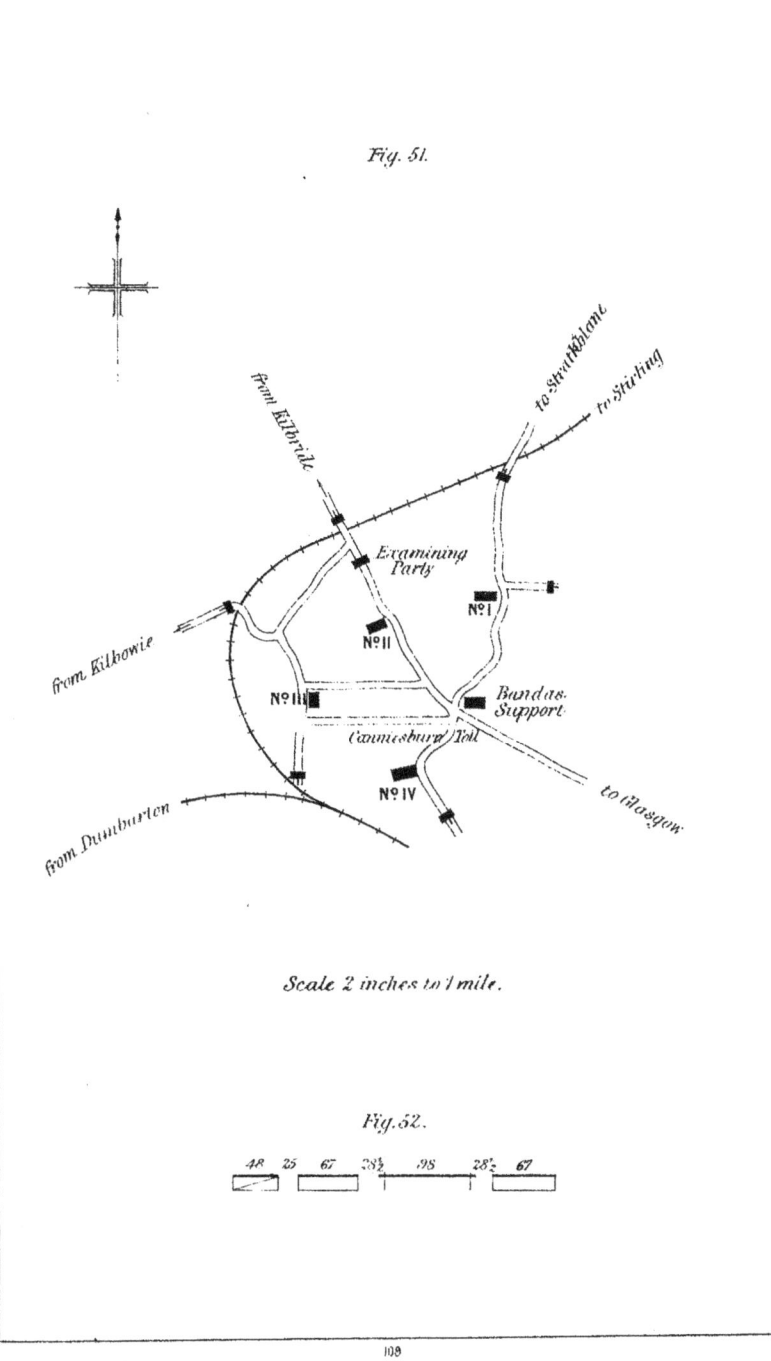

should be fed at one time. Saddles should be removed only once in 12 hours, bridles only for watering and feeding. A few horses only should be watered at a time, if the water is at a distance.

Vedettes should be placed from 20 to 30 yards apart, near enough to communicate when required, but not to carry on conversation. They should have their carbines drawn, loaded, and at the "advance." Lancers should remove or furl their lance flags. Vedettes.

A cavalry piquet should have ground in front of it suitable for aggressive action if called upon.

Cavalry would only be employed in support when the ground is very open, and suitable for their action. Usually the main brunt of resistance falls on the infantry. If there are cavalry supports, there should be one to every two or three cavalry piquets. A cavalry reserve is from 1,200 to 2,000 yards in rear of the supports. When cavalry forms the support.

The best time for practising outposts is in summer, though in favourable weather instruction may be combined with winter route-marching. If but few men are on parade, they should represent a reasonable part of the whole, the rest being imagined; and when limited to roads, the occupation of the intervening country must be taken into account (*vide* Fig. 51). * Practice of outpost duty to be as like reality as possible.

Officers and N. C. officers ought first to be allowed to complete and explain their arrangements, and afterwards such modifications as appear necessary should be pointed out.

CHAPTER VII.

MARCHES.

<small>Administrative considerations in peace.</small> IN peace time, or when not near the enemy, administrative considerations are allowed to preponderate, and the comfort of the troops may be studied. With several roads at disposal, we should give the infantry the shortest, cavalry the softest, guns the hardest and smoothest. Having only one road available, cavalry and artillery should march an hour or two before the infantry; first, to avoid their being fatigued by conforming to the slower pace of the latter, and secondly, because they have much more labour at the conclusion of the march.

<small>Tactical considerations at other times.</small> But in the vicinity of the enemy tactical considerations are paramount, and the "order of march" must be such as can be rapidly converted into an "order of battle."

<small>Advantage of a well regulated "order of march."</small> A well regulated order of march greatly conduces to the success of military operations. No obstacle, short of sheer impossibility, should prevent a commander arriving at the destination ordered at the time appointed.

A march must be carried out with celerity, and troops should be efficient at the end of it. This result can only be obtained by training, and reducing the weight the soldier has to carry. The pace of a long column is irregular, fatiguing, and more tedious. Hence, when practicable, a large force marches by several roads, communication being maintained between them; and the arrangements must be such as to insure the different columns reaching certain fixed positions simultaneously.

<small>Necessity for training and reducing weight to be carried. Use several roads when possible for a large force.</small>

Approaching the enemy, the different arms march in the order they will probably have to come into action—*e.g.* the guns near the head of the column, and cavalry in rear, although it is so harassing for them to regulate their pace by that of the infantry.

The following rules for arranging a march should always be borne in mind, viz:—

i. Never parade a moment sooner than necessary; and troops should be allowed to join the line of march direct from their camps.

<small>Rules for arranging a march.</small>

ii. When a column on the march has to retrace its steps, "turn about," and do not countermarch.

iii. If possible, avoid night marches.

iv. Take care that two columns do not cross each other on the march.

The *Field Exercises* (Part vii.) give clear directions for the conduct of route-marching. They are based on the orders framed by General Crawfurd for his celebrated Light Division in the Peninsula War.

<small>* Route-marching.</small>

As a rule, infantry for comfort and convenience march in fours, and observe the "rule of the road." But crossing open ground, or on very wide roads, such as are common on the Continent, the front may be increased to column of companies, half-companies,

<small>* Infantry usually march in fours, sometimes on a wider front.</small>

or sections. In certain circumstances infantry may also move in mass of quarter columns, or in line.

<small>Baggage on the line of march.</small> * The baggage of a single battalion should follow it; but in brigade the baggage of the different battalions follows in rear of the brigade in the same order as battalions stand in column. They must never march in the intervals between battalions. The only vehicles allowed immediately to accompany the troops are a water-cart and ammunition-cart per battalion, with perhaps an ambulance as may be ordered.

<small>Discipline to be enforced.</small> * The "line of march" is the occasion, above all others, when strict discipline must be enforced; and it is in movements by rail, road, and water, <small>Falling out should not be necessary.</small> * that good conduct is most conspicuous. No man may fall out without permission; and as halts are ordered after the first half-hour, and every subsequent hour, there ought to be no reason for men to leave the ranks.

<small>Object of intervals.</small> * Intervals are necessary between the different units to absorb accidental checks, and to avoid stepping out or closing up. But opening out, which troops are liable to do especially in warm weather, must not be allowed. With good discipline this tailing off should not exceed more than $\frac{1}{3}$ the length of the column. But with indifferent troops, and careless officers, it will be much more—perhaps doubling <small>Evil effects of tailing off.</small> * that length. And this is a matter which may be fraught with evil consequences; for a division, confined to a single road, should occupy 7 or 8 miles, and will require ordinarily about 2 hours to form for action; but if, owing to extreme opening out, it takes an hour or two more to get the rear

of the column into position, and the enemy meanwhile attacks in earnest, there is serious risk of defeat in detail.

Each company must conform to any deviation of the preceding company. When halted, the head of each battalion stands fast, and the companies in rear will close up when completely formed, unless a movement to a flank is contemplated. _{* Troops conform to the head of the column.}

No battalion, company, or section is ever to defile, diminish its front, or attempt to avoid a bad spot on the road, unless the preceding battalion or company has done so. On coming to a stream, or bad place, the men are not to be allowed to deviate from the direct course, although it may be convenient for them individually to do so. Officers must be on the alert to prevent this. Captains will go to the heads of their companies, and field officers to the heads of their respective half-battalions, and remain till their commands have passed, to see orders obeyed. _{* Defiling, &c., forbidden unless preceding troops have done so.}

When absolutely necessary, defiling must be executed with order and precision, by proper words of command, preceded by "attention."

It is better to pass pools of water, ditches, and other obstacles with a full front; for by defiling or picking the way a battalion is delayed ten minutes, and consequently a brigade half an hour—sufficient to upset any previous calculation for the execution of a time-march, and this might affect the plan of operations. _{* Delay resulting from defiling.}

If delay unavoidably occurs, the head of a company, after passing the bad place, must step short till the rest close well up; it will then rejoin the

I

114

preceding company by an increase of pace, or on arriving at the next check.

<small>Connecting files in case of great intervals.</small> * When intervals are great, files should be sent on to keep up connection between companies and battalions.

If a certain company cannot keep up, word should be passed by supernumeraries to the head of the column, and the commanding officer will check the pace. The rule to remember is to keep companies intact.

<small>Matters influencing rate of marching.</small> * The rate of marching is influenced by training, discipline, numbers, weather, and roads. A company can move a certain number of miles more easily than a battalion, and a battalion than a brigade. The dust of large bodies of troops is oppressive.

<small>Fitting and repair of shoes and socks.</small> * Officers and N. C. officers must see that the men's shoes and socks fit well, and are in good repair. Shoes with low heels and broad soles are the best. The feet should be well washed, and the bottom of <small>Remedy for sore feet.</small> * the socks soaped before marching. A good remedy for sore feet, mentioned in the *Soldiers' Pocket Book*, is to bathe them at night in tepid water, with a little alum dissolved in it; and to rub the tender parts with soft soap, or any sort of grease. Spirits and water applied to the feet is the best preventive against blisters. If, however, blisters rise, they should be pricked, but the skin must not be torn off.

<small>Breakfast before starting.</small> * The men should breakfast before starting. But the hour of starting depends on the season, climate, <small>Hour for starting.</small> * and distance to be got over. 5 or 6 A.M. is usual. But by the official account of Sir F. Roberts's march from Kabul to Kandahar, it appears that for the

first few stages the troops turned out at 2.45 A.M., and marched at 4.30. Later, when the thermometer stood at 100° in the shade at 4 P.M., they started earlier, at 2.30 A.M.

It is usual to halt for 10 minutes after the first half hour. A little longer halt may be made about midway, when the march exceeds 10 miles. If the column is long, a simultaneous halt should be arranged by signal or otherwise. In an ordinary march the advanced and rear guard calls are sounded on the bugle, and then the halt. In order to give the men the full benefit of the rest, there should be no needless closing up or dressing. * Periodical halts.

Troops must not be halted in villages, nor in a defile, if in the vicinity of the enemy. Not in villages or defiles near the enemy.

By column of route is meant a formation moving on a road with a narrow front on the line of march. * Column of route.

With infantry it is column of fours, the length of which is equal to the front in line, or twice as many feet as there are files. With a battalion of 8 companies of 100 men each, this will be 275 yards. * Length of a battalion in fours.

Cavalry usually move in column of route by sections, four abreast, the length of which is the same number of yards as there are horses in the ranks; in the case of a regiment of 4 squadrons of 100 sabres each this will be 400 yards. Or, they may be in columns of half-sections, two abreast, when the length of the regiment will be 800 yards. In neither sections nor half-sections are squadron intervals maintained. * Of a cavalry regiment in sections and half sections.

The usual column of route for artillery has a front of one carriage. The length of a field battery, with or without waggons, is calculated at 19 yards * Of a field battery with waggons.

Of a field battery without waggons. * for each vehicle, less 4 yards, the last carriage being taken as 15 yards. Thus the length of a field battery with waggons will be $19 \times 12 - 4 = 224$ yards; without waggons, $19 \times 6 - 4 = 110$ yards.

In a battery of horse artillery, each gun detachment consists of 10 mounted men. Each horse's length is 8 feet, and the distance between horses 4 yards = 12 feet. Therefore, each detachment, two abreast, requires $5 \times 4 = 20$ yards.

Of a horse artillery battery without waggons. Distances. * The length of a battery without waggons will then be $6 \times 19 - 4 + 20 \times 6 = 230$ yards.

The distances on the march between battalions is at least 30 paces to allow room for the bands, &c.

The distance maintained between cavalry and the other arms is 25 yards.

The distances on the march between batteries = half a subdivision, interval $9\frac{1}{2}$ yards; between carriages, 4 yards; between files, 4 feet. The distance between artillery and the other arms is $1\frac{1}{2}$ gun intervals, or as is usual with 6 horses, $28\frac{1}{2}$ yards.

These distances, however, may be relaxed ascending or descending hills: and if the width of the road permit, carriages may move up alongside each other to lock or unlock the wheels with the dragshoe.

Ordinary rate of infantry. * Infantry at the ordinary pace moving in small bodies, including slight checks, but not halts, marches at the rate of 3 miles an hour = 88 yards a minute.

The paces of the mounted services are, at a walk, 4 miles; at a trot, 8 miles; at a gallop, 12 miles an hour.

The rate of march of cavalry and horse artillery on fair roads, alternately trotting and walking, is 5 miles an hour = 146 yards a minute. *Ordinary rate of cavalry and horse artillery.*

The rate of field artillery may be taken, under the same circumstances, at 4 miles an hour = 117 yards a minute. *Ordinary rate of field artillery.*

Small bodies of the three arms combined, if unencumbered and on good roads, may be depended on to move at the ordinary infantry rate—3 miles an hour. But a division of all arms complete, even on fair roads, will not march faster than $2\frac{1}{4}$ to $2\frac{3}{4}$ miles: and an army corps than $1\frac{1}{2}$ to 2 miles an hour. *Ordinary rate of a division of all arms.*

If, for example, there is a small force consisting of— *Example of a small mixed force in column of route.*

    2 companies of infantry, each 100 strong in fours;

    A field battery without waggons;

    2 companies of infantry, each 100 strong in fours;

    1 squadron, 96 horses, in sections,

the length of the column of route is found as follows:—

2 companies in fours $\frac{200'}{3}$	= 67	yards.
Distance	$28\frac{1}{2}$	,,
Battery, $19 \times 6 - 4$	= 110	,,
Distance	$28\frac{1}{2}$	,,
2 companies in fours $\frac{200'}{3}$	= 67	,,
Distance	25	,,
1 squadron in sections 96 horses =	96	,,
Total length of column	422 yards.	

The front of this force drawn up in line will be 314 yards (*vide* Fig. 52). It will take $\frac{422}{88} = 5$ minutes to pass a certain point. And if it is ordered *Its front in line.*

Time required on the march and for deployment. * to move to a place 12 miles and 1,080 yards distant, and then deploy by 10 A.M., assuming the pace throughout to be 3 miles an hour, when should it march off?

The distance, 12 miles and 1,080 yards, requires 4 hours, 12 min.
Add 5 minutes for the last section to move off the ground, and 5 more for the rear to deploy } 10 ,,

4 hours, 22 min.

The troops should, therefore, start at 38 minutes past 5 A.M.

Punctuality. * Punctuality in concentrating a body of troops in a certain position is most essential. It is equally a mistake to arrive too soon as too late. Time-marches should be practised, the arrangements being left to the officers commanding the different units.

Time march of two battalions. * Thus, suppose two battalions are ordered to arrive at a village by 12 noon by two different roads, found on a map to be 6 miles 1,080 yards, and 8 miles 700 yards distant respectively, when should each march off?

6 miles takes 2 hours, $\frac{1,080}{88} = 12$ minutes; therefore the first battalion requires 2 hours 12 minutes, and should move off at 9.48 A.M.

8 miles takes 2 hours 40 minutes, $\frac{700}{88} = 8$ minutes; therefore the second battalion requires 2 hours 48 minutes, and should move at 9.12 A.M., or 36 minutes before the first battalion.

Time march of two companies. * A company is ordered to a cross road on a moor, distant by road 5 miles 540 yards; a second company is directed to the crest of a hill, distant 6 miles 450 yards, at what hour should each march, so as to arrive simultaneously at 10 A.M.?

5 miles takes 1 hour 40 minutes, $\frac{540}{88} = 6$ minutes, therefore the first company requires 1 hour 46 minutes; and should march at 8.14 A.M.

6 miles takes 2 hours, $\frac{450}{88} = 5$ minutes, therefore the second company requires 2 hours and 5 minutes, and should march at 7.55 A.M.

Again, two troops of cavalry are ordered to rendezvous by 10 A.M. by different roads, found on the map to be 8 miles 1,600 yards, and 10 miles 730 yards long respectively, when should each start? The speed is taken at 5 miles an hour.

Time march of two troops.

8 miles requires 1 hour 36 minutes, $\frac{1,600}{146} = 11$ minutes, therefore the first troop requires 1 hour 47 minutes, and should march at 8.13 A.M.

10 miles requires 2 hours, $\frac{730}{146} = 5$ minutes, therefore the second troop requires 2 hours 5 minutes, and should march at 7.55 A.M.

If an infantry brigade (3 battalions of 1,000 rank and file each), 1 field battery, and 2 squadrons of cavalry each 54 files, are ordered to a camping ground 12 miles distant, and there are three available good roads, of which one is allotted to each arm, when should the troops march off, so as to arrive at noon?

Time march of a mixed brigade by three roads.

The infantry, at 3 miles an hour, requires 4 hours, marches at 8 A.M.

The artillery, at 4 miles an hour, requires 3 hours, marches at 9 A.M.

The cavalry, at 5 miles an hour, requires 2 hours 24 minutes, marches at 9.36 A.M.

If the whole were limited to one road, the rate will be governed by the infantry, making the time 4 hours, and the troops must march at 8 A.M.

<small>And by one road.</small>

* A flank march occurs when one flank is exposed to attack, as when an army is marching in a certain direction, and the enemy is moving perpendicularly to the line of march. It is dangerous only when the enemy is within striking distance. To be successful the troops must be highly trained and well led. The commander will doubtless have had ample warning, and can arrange his order of march accordingly. The march must be completed rapidly, and yet the troops must be always on the alert to receive an attack in front and in flank. Guns should be distributed between the front and rear of the column. The main body of the cavalry should be on the side furthest from the enemy, lest, if on the other side, they are driven in disorder upon the column before it can form up. But part of the cavalry must reconnoitre the whole exposed flank, as well as the front, scouting further towards the enemy if the country is open; and a body of infantry should always support them. Any defiles on the exposed flank must be explored, and guarded if possible. Every precaution should be taken to conceal the march by keeping woods, marshes, streams, ravines, and other features of the ground on the side nearest the enemy.

<small>Meaning of a flank march.</small>

<small>When dangerous.</small>

<small>Dispositions.</small>

* A night march should not be undertaken unless absolutely necessary; for the direction may be lost, troops get into disorder, and are done up for the
* following day. Landmarks ought to be fixed beforehand along the routes; a guide should lead every

<small>A night march.</small>

<small>Guides and landmarks.</small>

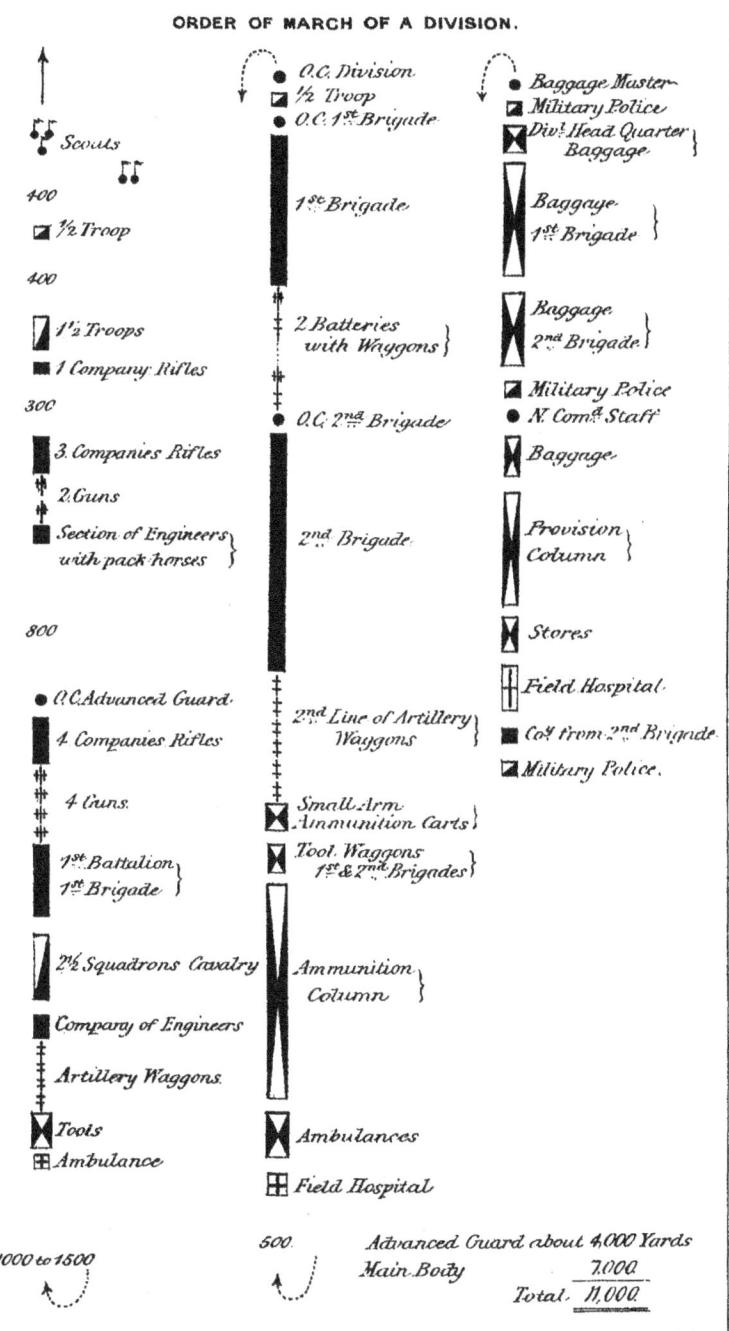

column; and intelligent men should be left by the advanced troops at cross-roads to indicate the way to the rest. The bayonet alone should be used in a night action, and no firing. The successful attack on Tel-el-Kebir, before daylight, is fresh in memory. * The enemy's position was not shaken by artillery fire, and the Egyptians were surprised, having no outposts. How the late Commander Rawson, R.N., guided the attacking columns by the stars, and the wonderful discipline and steadiness of the troops on that occasion, are matters of history.

<small>* Tel-el-Kebir.</small>

Some idea of the composition, strength, and extent on one road of an infantry division may be obtained by reference to Plate VI., which represents the order of march recommended by the quartermaster general.

<small>Order of march of an infantry division.</small>

Allowing for opening out half a mile between the rear of the advanced guard and the head of the column, and between the rear of the column and the head of the baggage convoy, the whole covers from 7 to 8 miles; which is one basis of rough calculation of the time required for the advanced guard to "contain" an enemy, to enable the whole to take position in line of battle.

A day's march is usually from 12 to 15 miles, with a halt every fourth or fifth day. From 15 to 20 miles a day can be done exceptionally. Beyond 20 miles a day would be considered forced marching. But in 1813, for three days previous to the battle of Leipsic, the French army marched 30 miles a day.

<small>* An ordinary day's march.</small>

<small>Often exceeded.
* French in 1813 before Leipsic.</small>

Bad roads, however, make a vast difference. Thus,

Effect of bad roads, e.g. Blucher, Wavre to Waterloo, 1815.

* Marshal Blucher, hurrying the Prussian army from Wavre to our assistance at Waterloo in 1815, could only move 1½ miles an hour.

Road blocked with baggage—Canrobert at Magenta 1859.

* Again, when a good road was blocked with baggage, Marshal Canrobert's corps advancing to support the French guard at the battle of Magenta in 1859, took 5 hours to accomplish 9 miles.

Some notable marches.

* Some notable marches have taken place; and, to show what can be achieved, a few are briefly referred to. They should be carefully read in history.

Napoleon from Boulogne to the Rhine, 1809.

* In September, 1809, Napoleon marched from Boulogne to the Rhine with 150,000 infantry. He was joined by cavalry and artillery on the way. The distance marched was 400 miles in 25 days, at the rate of 20 miles a day, or, including halts, 16 miles. His object was to surprise the Austrian army, and he encountered no opposition. But this remarkable march occurred before telegraphs were invented and intelligence could be rapidly transmitted.

In Alison's *History of Europe* it is stated that the instructions given by Napoleon to all the chiefs of the grand army for the tracing of all their routes and the regulation of their movements were as perfect a model of the combinations of the general as the fidelity and accuracy with which they were followed were of the discipline and efficiency of his soldiers. The stages and places of rest, the daily marches of every regiment, were pointed out with undeviating accuracy over the immense distance from Cherbourg to Homburg.

Sherman from Atlanta to Savannah, 1863.

* In November, 1863, the American General Sherman marched from Atlanta, the capital of the State of Georgia, to Savannah, a port on the Atlantic, 250 miles, with 60,000 infantry, 5,500 cavalry, and 60 guns, meeting constant opposition and obstruction, in 27 days. "All the troops were provided with good waggon trains, loaded with ammunition and supplies —about 20 days' bread, 40 days' sugar and coffee, a double allowance of salt for 40 days, beef cattle for 40 days' supply. The waggons also carried 3 days' forage and grain. All were instructed by a judicious system of foraging to maintain this order of things as long as possible, living chiefly, if not solely, on the country."

In August, 1880, General Sir F. Roberts, with 7,500 infantry, 1,600 cavalry, and 18 mountain 7-pounder guns, and 8,000 camp-followers—total, 18,000 men and nearly 9,000 animals—marched from Kabul to Kandahar, 321 miles, in 23 days, at an actual rate of 15⅓ miles a day, or, including halts, 14 miles. No wheeled guns or transport accompanied, but everything was carried on pack animals. There was no opposition. * Sir F. Roberts from Kabul to Kandahar, 1880.

Lastly, in Napier's *History of the Peninsula* we read that on the 29th July, 1809, Brigadier R. Crawfurd, with the 43rd, 52nd, and 95th, reached Wellington's camp after the battle of Talavera, and immediately took charge of the outposts. After a march of 20 miles, Crawfurd, fearing for the army, allowed only a few hours' rest, and then, withdrawing about 50 of the weakest from the ranks, recommenced his march with a resolution not to halt until the field of battle was reached. Leaving only 17 stragglers behind, the brigade crossed the field of * Crawfurd's light division to Talavera, 1809.

battle in a close and compact body, having in that time passed over 62 English miles in the hottest season of the year, the men carrying from 50 to 60 lbs. weight upon their shoulders.

Transport by Railway.

Notice to station-master. * The regulations are laid down in the last pages of the *Field Exercises.* Sufficient notice should be given to the station-master. Women and children must be at the station half an hour before the departure of the train, conveyed in cabs, or army service waggons if available, in charge of a N. C. officer provided with a nominal roll. He will see them at once into the carriages. Officers' light baggage, with the servants as a guard, should arrive at the same time. After placing the baggage in the train, the servants enter the carriages allotted to them.

Married families. *

Light baggage. *

An officer or N. C. officer will go before, and with the station-master chalk the letters of the companies on the footboards of the carriages. In the case of a battalion, the adjutant should precede by about ten minutes, accompanied by a marker from each company, one for the band, and one for the guard, whom he will place along the platform, standing beside the train, abreast of the compartment where the head of each company will rest. These N. C. officers should have a note of the exact numbers proceeding.

Adjutant and markers. *

Capacity of compartments. * As a rule, a compartment intended for 10 ordinary passengers will be allotted to 8 soldiers ; and one for 8 ordinary passengers to 6 soldiers.

Each company, arriving at its marker, will be

halted, and ordered to face the train. Arms will be grounded. The men nearest the train must be careful that the muzzles of the rifles do not project beyond the platform. At the command, " Take off valises," one man of each compartment enters and places his valise under the seat, and his rifle on the seat; then he takes, and places similarly, those of the other men in regular order, so that each man's valise is under his own seat. Haversacks and water-bottles are turned round to the front. The men take up arms and enter the carriages. Each man retains hold of his rifle, unless the commanding officer allows them to be placed over the valises. But they are never to be placed on the floor. * Embarkation.

Perfect silence must be maintained. No man is to put his head out of window, nor is he to leave his carriage without permission.

On arrival at the destination, the men get out with their rifles, and fall in as when they embarked. The last man in the compartment hands out the valises. Arms are grounded, accoutrements and haversacks adjusted, and the valises put on. Arms are then taken up, and the companies marched out of the station in their original order. Or the commanding officer may order the markers to be placed in a convenient space. * Disembarkation.

CHAPTER VIII.

MINOR OPERATIONS.

CERTAIN operations in war require some special notice. These are the conduct and attack of convoys; the defence and attack of houses, villages, and woods; the defence and passage of a defile, and of a river.

Conduct of a convoy.   A convoy is an attendant force on the march, *i.e.* the escort required for the defence of transport other than the train accompanying an army. But the term is generally applied to the transport itself. It may be either moving from the base, in the same direction as the army, carrying provisions and clothing; or from the front to the rear with sick, wounded, and prisoners, no single line of waggons should cover more than a mile of road. If there are more carriages, divide them into detachments, and send them as different convoys.

The principle is on the march to afford the greatest amount of safety to what is most valuable, and to place the pack animals in front, for wheels cut up the road.

If the road is wide enough for 3 carriages, and the order of march is likely to be undisturbed for an hour, move two carriages abreast. But if the road narrows frequently, passing villages, and crossing streams, adhere to single file. The regulation 4 yards distance should be maintained between carriages, except when ascending hills. If the road is steep, double-horse each carriage, and ascend in two divisions. To avoid accidents and overloading in consequence, a percentage of spare carriages, wheels, poles, &c., should be provided.

Short halts should be made at regular intervals to rest the horses, and close up the column. On a long journey the convoy should be parked for feeding in a spot convenient for moving on. The horses must not be taken out of the shafts, but fed with forage carried for them on the waggons.

The value of the convoy, the length of the journey, and the friendliness or otherwise of the country, determine the strength and composition of the escort. <small>Considerations determining strength and composition of escort.</small>

As a rule, it is chiefly infantry, with enough cavalry to reconnoitre, and a party of engineers to remove obstacles, and repair passages and bad parts of the road. Guns are sometimes added. The escort is necessarily disseminated, because of the length and weakness of the convoy; but it should not be needlessly divided, and a force must be kept ready to meet the chief attack.

There should be an advanced and rear-guard and flankers, and small parties detached from the main body to the head and tail of the column. Cavalry are best suited for the advanced guard, to explore the country and discover ambuscades and the direction of any impending attack. <small>Disposition of escort.</small>

Their patrols should reconnoitre at least 5 miles to the front and flanks. This must be specially done before starting after halting at night.

* If the leading waggons are stopped there will be a block; therefore, some infantry should march at the head to check a dash made by cavalry. A similar party should be at the rear. The usual position for the main body is central, when the convoy does not exceed a mile in length. But in an inclosed country it is better at the end which is most threatened, and it may detach a small party to the centre (*vide* Fig. 53).

Escort of prisoners.

* The usual strength for an escort of prisoners is 10 infantry and 1 cavalry soldier to every 100 prisoners, occasionally with guns added. The escort is more subdivided to suppress insubordination. When halted for the night, guards must be mounted.

Duty of escort of prisoners when attacked.

* If a convoy of prisoners is attacked, the escort should remain close to them, as the enemy will not risk killing their own people by firing. To prevent escape the prisoners should be ordered to lie down.

Arrangement of a mixed convoy.

In the case of a mixed convoy, consisting of ammunition, provisions, *matériel* and clothing, wounded and prisoners, the arrangement on the march should be first the ammunition, then the provisions, last of all, the effects and clothing.

Information before starting.

The officer commanding before starting should obtain information of the road, places for halting and billeting, where the enemy is, where he may expect attack, and where he may take refuge in case of retreat.

A 4-horse waggon occupies 12 yards in the convoy.

The rate of travel may be taken on a good road

Fig 53.

CONDUCT OF A CONVOY.

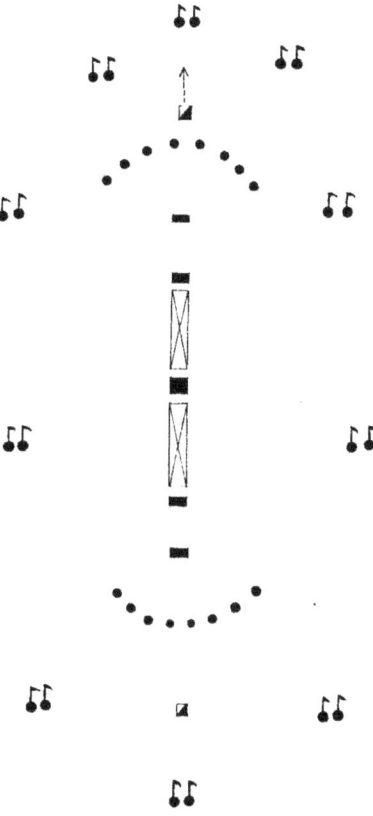

at $2\frac{1}{2}$ miles; in a hilly country at $1\frac{1}{2}$ miles an hour.   *Rate of march.*

Every precaution must be taken against surprise. Towns, villages, and defiles should be avoided.

On coming to a river or bridge, supposing the rear to be secure, the reserve should cross first; if it is not, the reserve should follow. If both front and rear are threatened, the main body may be divided.   *Precautions against surprise. Crossing a river or bridge.*

Detachments should be sent on from the reserve to occupy a defile. The convoy must close up on the widest front. Advanced and rear guards are withdrawn. Then the defile is reconnoitred. The advanced guard first passes, and proceeds far enough to admit of the convoy being parked in rear beyond the defile. The reserve occupies the high ground in the direction of the enemy.   *Passing a defile.*

If the convoy consists of powder and combustibles, let nothing else be placed in the carriages, and no one must ride on them. Forbid smoking. Do not enter villages, and, unless specially ordered, move only at a walk.   *Precautions with powder, &c.*

An intelligent soldier should be told off to each carriage, or if the escort is weak, to every 2 or 3 carriages; or mounted men may be distributed to superintend and see that the column is well closed up and the drivers obey. Prompt obedience must be insisted on from the drivers, but neither they nor their horses should suffer ill-treatment.   *Duties of men accompanying the carriages.*

About $\frac{3}{4}$ hour after starting, halt for about 10 minutes, and afterwards at similar intervals. But take care to halt clear of bridges and all obstacles.   *Halts*

If the drivers are civilians, and are specially

K

requisitioned, they must be well watched, as in the confusion of an attack they may cut the traces and desert with their horses.

As a rule avoid fighting. All needless fighting should be avoided. But if the enemy has occupied defiles, or commanding ground, endangering the march, he should be attacked with all the force that can be spared to dislodge him, detachments being left with the convoy, which should continue to advance unless the road is stopped. When the enemy is dispersed, he should not be pursued.

How best to meet a superior force. If a superior force of the enemy is reported, it is best to halt, and park the carriages, if possible, in some inclosure near the road, forming an intrenchment, and to collect the escort. But if it is impossible to quit the road, the carriages should be closed up in double files, and part of the escort sent forward to delay the enemy.

Distinguish between a real and a false attack. If the enemy makes two separate attacks, a real attack should be distinguished from a false attack; and if the escort is not strong enough to meet both, unite it, and first fall on one attack, then on the other. On no account is a detachment posted to defend one part of the convoy to quit its post and assist another, without distinct orders.

Conduct if escort is defeated. If the attack succeeds, try to get away with part of the convoy. But if none can be saved, endeavour to save the horses.

When the head of the column is attacked, maintain order, and turn the carriages carefully, lest the road be blocked.

If pursued closely, upset a couple of waggons. If attacked by surprise by a strong force in flank, concentrate the escort, and do the best to save part.

When the enemy is not reported in superior force, it will not be necessary to park, nor even to halt. The escort should take up a position covering the flank threatened, and move along with the convoy.

When the enemy is not superior.

If the enemy is inferior in strength, part of the escort should advance, and disperse them.

When he is inferior.

When obliged to halt on the near side of a defile, while it is being reconnoitred, it is well to park the leading half of the convoy, which may feed and rest, while the second half passes through and parks on the other side. The other half will then follow.

The best way to park simply for a halt, is to form the waggons in several lines, 25 yards apart, axle-tree to axle-tree, poles and shafts in the same direction; the horses being picketed in front of their respective waggons.

* Parking for a halt.

But in order to resist an attack, draw up the waggons in a square or oval, in one rank; or better in two ranks, if there is sufficient interior space. When the waggons are placed axle-tree to axle-tree the barrier will be stronger; but when end to end, they will give more space. The horses are picketed inside, opposite the waggons. The men of the escort can fire from the corners, and between the waggons.

* To resist attack.

When ammunition is parked, the carriages should be massed close together, and the escort should take up a defensive position at a distance, to prevent the enemy firing on the convoy. Ammunition must be defended to the last; and everything which cannot be saved should be destroyed.

Should there be no time for parking when

When pressed for time.	* attacked, close up the waggons in double files, and turn the horses inwards, to shelter from cavalry.
Dispositions when halted.	* When halting for the night, choose a defensible position, in an uninhabited place, with good water near. Post piquets and send out patrols, and make every disposition to prevent surprise.
Convoy by water.	* When a convoy is in boats, one half the escort should be in the boats; the other half on land, by the river or canal bank, disposed much in the same way as for a convoy moving by land. They must reconnoitre well, and protect the convoy. In case of attack they will be reinforced by the men in the boats.
Infantry usually in boats, cavalry on land.	* Infantry will generally be in the boats, while the cavalry on land reconnoitre; and if this duty is efficiently performed, so that timely notice of an enemy is given, the infantry can be easily landed. Whereas, if the infantry march along the banks, the men are fatigued, and the convoy retarded.
Sometimes infantry on land.	* But ascending streams, the progress is slower, and it is advisable for some infantry to march on land to protect the horses and drivers.
Connecting files.	* Connecting files should pass intelligence between the farthest patrols and the river.
Conduct when attacked.	Should the convoy be attacked by water, if there are any guns, they should fire on the enemy's boats, and the infantry fire on his men.

If the enemy is signalled approaching by land, collect and draw the convoy to the opposite side. If the attack be made, halt, and land the escort, keeping the boats ready for re-embarkation.

Should the escort be defeated, part of the convoy

should endeavour to escape at the best. The boats should be sunk rather than surrender. When possible, let the escort retire fighting, then take to the boats, and rejoin the convoy. When attacked on both banks, the escort must be divided.

In case of defeat.

The best way to attack a convoy by water is by ambush, and at favourable points on the river. Ascertain when it is likely to pass, and act secretly. Fire with artillery on the leading boats, and with musketry on the boatmen. If the convoy is not brought to a standstill, bring all available men into action, follow the convoy, and keep up the fire. If the leading boats are disabled or surrender, the rest will probably do the same.

Attack of a convoy by water.

An attack on a convoy on land will have greatest chance of success if made while the convoy is moving through difficult ground, or a defile; when parking after a march, or starting after a halt. It ought to be a surprise, so as to give the column no time to close up, or the escort to concentrate. It will be more difficult if the convoy reconnoitres well. The most disadvantageous spot to the convoy should be chosen, or an ambush may be laid. The attack should be vigorously made on the flanks in an open country; on the extremities in an inclosed country; or, let part of the convoy enter a defile, or part issue from it, and then attack. Cavalry is chiefly employed. Perhaps ⅔ the attacking force will be of this arm, with the remaining ⅓ infantry, and some guns added.

Attack of a convoy on land.

If the attacking force is superior, it may fall at once on the head, rear, and centre; or part may

When attacking force is superior.

head the convoy, while the main body defeats the escort in detail.

When it is inferior. But if the force is very inferior, it is best only to make an attempt on the rear, and to hang on the line of march to delay and harass the convoy.

Without guns an attack on a convoy, closed up or parked, will not probably succeed, unless with a superior force.

It is advisable to feint at many points in order to disseminate the escort, and to have a reserve. If **Action if attack is successful.** the attack succeeds, move off the carriages quickly; and in case there is danger of recapture, remove what is most valuable. If nothing can be removed, take away the horses, and burn the convoy.

Escort to foraging parties. * Sometimes foraging parties require an escort, and its strength and composition depend on the nature of the country, the distance to be proceeded, and the proximity of the enemy.

After clearing the country of the enemy, the escort should be disposed in a chain of posts, with a reserve in front of the district to be foraged. Piquets should be placed on the approaches, and infantry and cavalry patrols pushed on in front.

The posts should be so far to the front as will allow time for the foraging parties to mount, and support the escort, or retire in good order.

To forage a village. * In foraging a village, place a chain of sentries round it, to prevent all but the parties detailed from entering. If the inhabitants do not carry out the forage, parties must be sent in to collect it. An armed party should patrol round the village to prevent straggling and disorder.

Defence of a House.

A well-built house, capable of holding at least half a company, may be turned into a small fortification, without much labour, so as to afford cover and an obstacle. The selection is generally made in reference to an extended position, a detached post, a bridge, &c. (*vide* Fig. 54). Choice of a house.

It should be sheltered from artillery fire when possible. But this cannot always be, and a good defence can still be made if precautions are taken to put out fires caused by shells, and all inflammable substances removed. Shelter from artillery when possible.

A brick house with slated or tiled roof is the best; for stone splinters under artillery fire; and wooden or thatched houses easily catch fire. A flat roof enables the defenders to fire over the parapet wall by means of loopholes or sand-bags. Parts of the house should flank each other. For this object bay windows and porches are useful. Brick better than stone.

Any cellars give some accommodation, or else splinter-proofs may be built behind a second wall, for shells burst at the first wall. These will render men fairly safe, and protect them from falling splinters during the cannonade which precedes the advance of the enemy's infantry. Cellars.

The defensive measures to be taken are—the very first thing to clear the field of fire. * Defensive measures.

To loophole the outer walls, providing, when possible, flanking fire, and making two rows on the ground floor if not pressed for time. The loopholes of the upper rooms should be sloped well down to see as close to the foot of the wall as possible. Some * Clear field of fire.

Loopholes. * must be made at the salient angles. When walls are very thick, loopholes are more easily made under the eaves, where they are thinner. Low walls or flat roofs should be prepared for men kneeling. If the roof is high-pitched and tiled, the single tiles just above the eaves should be removed.

Barricade doors and windows. * Strongly barricade the doors and windows on the basement and ground floor to the height of about 6 feet, with boxes and casks filled with cinders, earth, or anything else handy; or by bullet-proof timber fastened by struts. Loophole the upper part of the doors and window-shutters. The men inside

Remove glass. * can use the furniture as a step. Remove all glass. The up stairs windows need not be barricaded, but sheets or blankets should be hung across the lower parts to hide the men from view. Rolls of carpets, mattrasses, sand-bags, &c., may be piled up 6 feet high, as a protection, with openings left to fire through.

Provide flank defence. * If there is no flank defence afforded by the building itself, build out a tambour, at an angle, on a projecting gallery, on the upper floor, from which to fire down.

Obstacles. * Ditches should be cut outside the lower windows and doors, and obstacles strewed round them to prevent the enemy from closing.

Communications. * Cut communications through partition walls, and leave one entrance on the least exposed side for communication or retreat. Loophole the porch inside the door (*vide* Fig. 55).

In preparing an obstinate defence, remove the staircases, and loophole the upper floors, communicating with the latter by ladders.

FORTIFIED HOUSE.

Fig. 54.

DEFENCE OF A HOUSE.

Fig. 55.

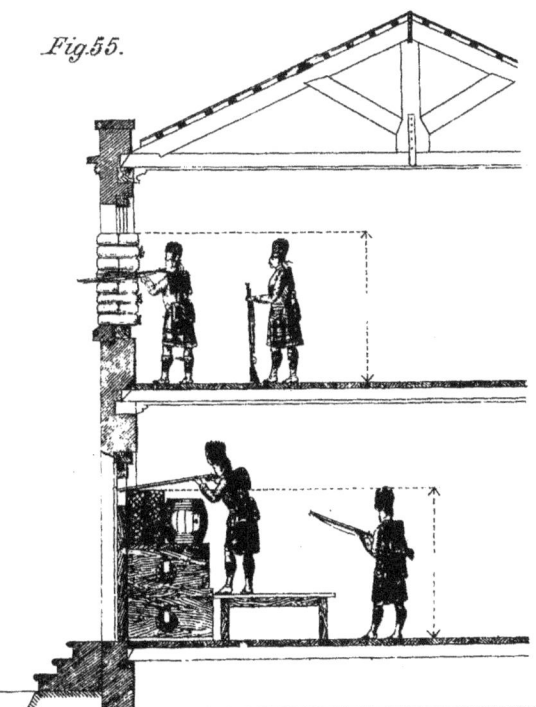

If breaching is feared, support the upper floor with beams, and have material ready to barricade the breach.

Provide against fire by removing thatched roofs. Cover the floors with wet earth or dung. Place barrels of water and wet blankets in each room, and detail a few men specially as firemen.

* Precautions against fire.

For the defence of a farm or a group of buildings, similar precautions must be taken on a larger scale. Boundary walls should be occupied, to obtain a good distribution of fire, attending to flank defences. Open as far as possible in the rear any small outhouses useless for defence. Loophole only the walls commanding the enemy's line of attack. What houses are occupied should flank each other, and see into the boundaries. Make the best communications between them, and arrange one as a keep. Such a farm or group of houses, holding 2 or 3 companies, often becomes a good tactical pivot.

* Defence of a farm or group of buildings.

Attack on a House.

Employ guns and rockets before the assault, to clear the way for the infantry by destroying obstacles, making a breach or setting a building on fire. If we have no artillery, at least two different attacks should be arranged on different sides, and an attempt made with ladders to get in by the roof. An extended line precedes the storming party, getting as close to the houses as any cover will allow, for the nearer they are the less will they be exposed to fire from loopholes of unflanked buildings, because of the difficulty of obliquing and depressing the rifles.

* Artillery.

* At least two attacks.

* Extended line.

They aim at the loopholes, unbarricaded windows, and other vulnerable parts. Under cover of this fire the storming parties advance, generally on the angles, and endeavour to blow in the doors or lower windows, or to force them open with crowbars, or a piece of heavy timber. An effort may be made to undermine the walls, while a ladder party tries the roof. Any thatched or wooden outhouses should be set on fire, and even a fire of straw or brushwood close to the walls will shake the confidence of the defenders. A reserve must be kept in hand to follow up success, to cover retreat, and to prevent aid being rendered to the defenders.

Storming party. *

Burn what is inflammable. *

Reserve.

Defence of Villages.

When villages are useful.

Villages are useful in a defensive line as supporting points or pivots; but we should compare the number of men required to hold a similar front in the open, and see if the garrison required for the village will be more than its position is worth.

They possess the advantages of being made defensible in the shortest time, and of being defended obstinately for a long time; they shelter the troops before an attack; and conceal their disposition, and those of troops in rear.

Disadvantages.

On the other hand, the defenders are scattered, and it is difficult to supervise them. Artillery fire causes loss by splinters, &c., and shells may set buildings on fire.

A village may be a pivot on the main line, an advanced post or outpost, or a reserve post; and the

Plate VII.

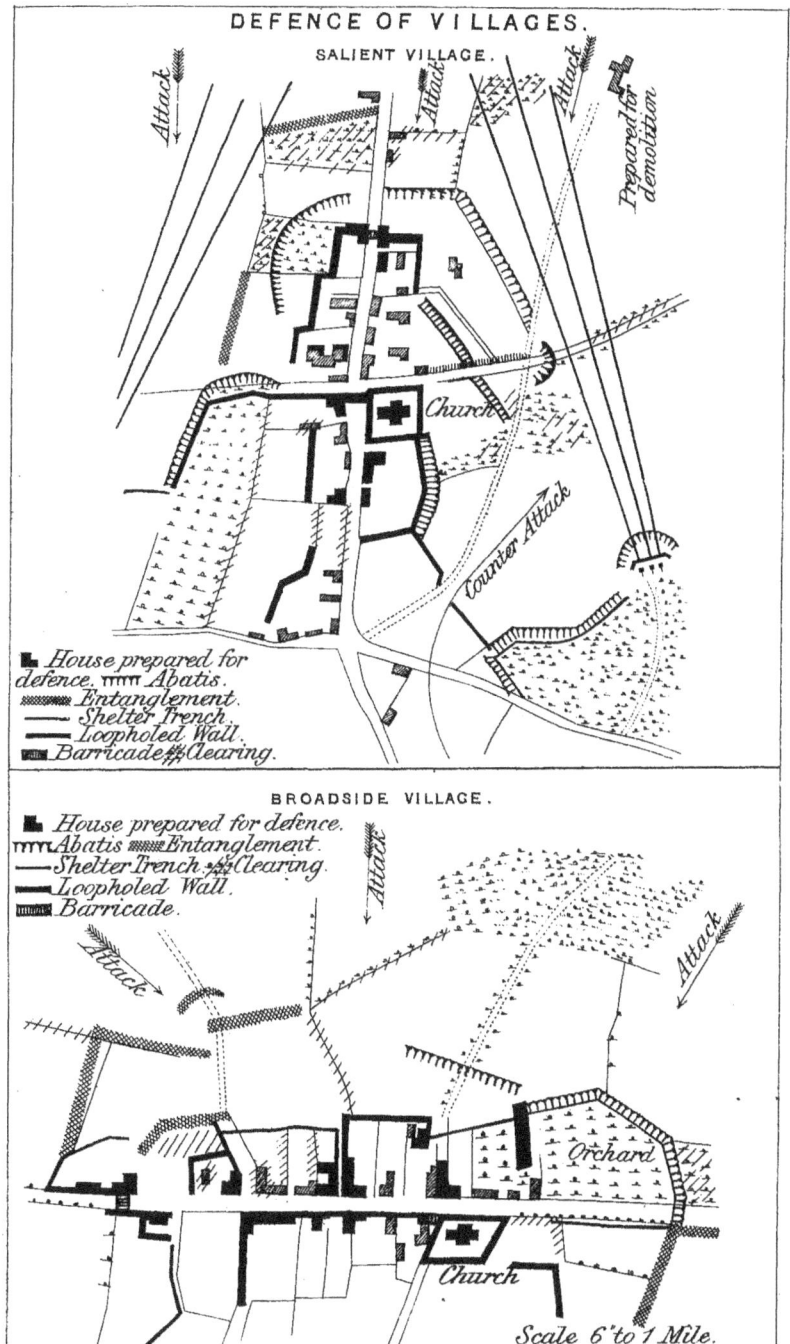

DEFENCE OF VILLAGES.

CIRCULAR VILLAGE.

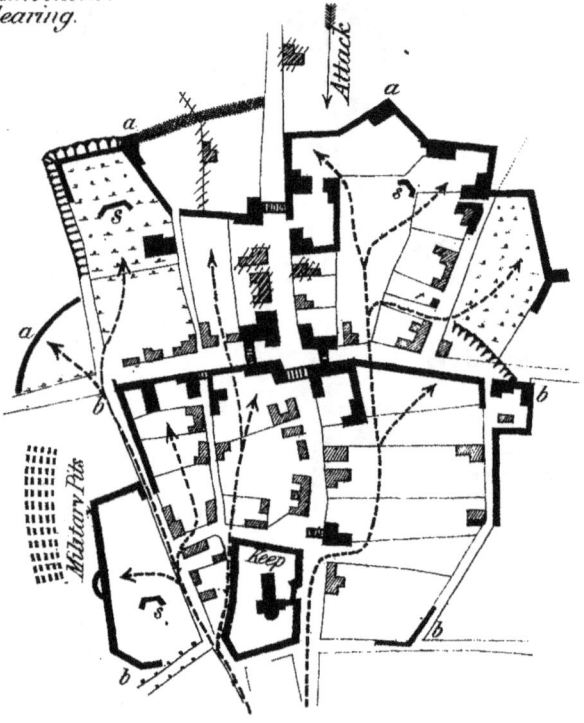

a, a, 1st Line of Defence.
b, b, 2nd do. do
■ House prepared for defence
s. s. s Support Trenches.
▬ Barricade
--→ Communication.
ⅿⅿⅿ Abatis.
▰▰▰ Entanglement
//// Demolition.
✶✶✶✶ Clearing.

following principles should be followed in strengthening it, viz.:—

To make the front and flanks strong, and the rear capable of resisting infantry, so that troops in reserve outside may have time to act. The distance in front of the general line will decide whether the defence is to be obstinate or not, and what extent is to be fortified. But if the village is in rear of the line, it must be obstinately defended all round.

<small>Principles of defence.</small>

There should be no high ground within close range, and the surrounding country should afford a clear field of fire with the aid of a little labour, and allow troops outside to advance unimpeded in support.

<small>Must not be commanded.</small>

A village may be either broadside to the enemy, salient, or circular (*vide* Plate VII. from the *Manual of Field Engineering*).

Form affects the time for preparation, and the obstinacy of defence. Thus a hamlet, end on to the enemy, may be made strong against flank attacks, but will be easily raked, and will require shelter trenches on each side to increase the front fire. A village broadside on is strong in front, and safe from fire, but the flanks must be seen to. A circular village is suitable anywhere.

<small>Preparation and defence depend on form.</small>

The increased range of artillery makes villages untenable at a greater distance now. Shells will set on fire hay ricks, thatched roofs, and wooden houses, and troops cannot remain in houses during a cannonade. Therefore, having regard to modern artillery, the arrangement should be somewhat as follows:—The extended or shooting line is placed outside the village, under cover of such inclosures, walls, hedges, and fences as are selected as suitable

<small>* Arrangements to meet artillery fire.</small>

<small>* Extended line.</small>

in form to the general line of defence, supplemented when needed by shelter trenches, and far enough advanced, at least 40 yards, from the houses to be clear of splinters; but 150 to 200 yards in front will be better unless the front be thereby too extended, so that if the enemy gain the first line, the defenders can fire into it from the loopholed houses.

Guns best behind epaulments. Sometimes a field work may be made, and guns placed in it. But guns are much better placed behind epaulments giving a cross-fire in front; and *Machine guns.* machine guns will be very useful in defence, being light and easily moved.

As obstacles, though they oppose any advance of the enemy, may prevent the defenders taking the *Wire entanglement.* * offensive, and afford shelter to the enemy, wire entanglements are the best, giving no cover, not affected by artillery fire, and entirely putting a stop to the action of cavalry.

No village should be held situated in a valley; but those built on high ground, or near a river or stream, become important and of great value to an army deploying.

We must, on first examination of a village, find out the nature of the surrounding country, cover available for the enemy, and whether the banks, fences, &c., can be defended; then determine quickly *Fix outer line first, then inner defences.* * what is to be the outer line, and where shelter trenches are required; then choose the best houses for the inner defence, avoiding wooden buildings which cannot be pulled down. Also, examine the roads traversing the village, fix places for barricades, and make any openings required. No more should be undertaken than we have time and means for. Begin the outer line first, and when this is nearly completed, devote the rest of the time to the other parts.

When a village is used as a detached post, it should be made secure all round. <small>As a detached post.</small>

When it is an advanced post, supported by a general line, it should be very strong on three sides. The rear should be open, and a line of shelter trench thrown up 250 yards in rear. This trench will prevent a flank attack of the enemy getting into the rear, and with gun epaulments stop him from using the village if captured. If the enemy takes any part by surprise, the moment of capture is the very time to drive him out before his men can get into order again. For this reason a reserve must be held in hand ready to act quickly. <small>As an advanced post.</small>

<small>* Necessity for reserve.</small>

The supports are best placed in small bodies clear of the village, and immediately in rear of the troops they are to support, protected by shelter trenches, or prepared fences if there is time. The troops should not be put into the loopholed houses until the last moment; the men told off to their places; and as soon as the enemy attacks, they should be brought into the village in small numbers, and posted by the officers. By this time the cannonade will have moderated. <small>* Best positions for supports.</small>

<small>* Troops not to enter the houses till the last moment.</small>

When the greater part of the defenders are in the village, troops of the 2nd line should line the shelter trench, and others may be conveniently extended in intermediate positions, so as to feed the fight, and gain the moral effect of moving forward. On no account should the defence be simply passive. <small>* No passive defence, but the fight to be fed.</small>

A well-posted extended line should hold the outer defences to the last, having supports and reserves ready to re-inforce; and easy communications should be arranged for their advance, and also for an orderly retreat.

Garrison. * A sufficient garrison would be of infantry, 2 men to each pace of the circumference, allowing for reserves; 1 man to a pace is the least. A few field guns and machine guns may be added, posted on high ground in rear, or to command roads; but the guns must be able to withdraw safely. When skilled labour and demolitions are necessary, a few engineers should be included.

How the troops should be told off. * The troops should be told off by companies, half battalions, or battalions to prepare and hold the different parts of the village. The shooting line and its supports will require $\frac{2}{3}$ to $\frac{3}{4}$ the garrison. It is desirable to prepare short walls or hedges perpendicular to the front, to guard against the enemy spreading right and left, should he succeed in breaking through; and to divide the front into a series of defensive sections, each to be manned by a company or half a company.

During the cannonade the men should be kept well under cover in trenches, or field casemates, or behind walls. In ordinary houses they will not be safe unless in cellars; though they will be fairly so behind houses, as shells will burst after passing through one wall.

A keep prepared when desperate resistance intended. It is useless to prepare a keep unless a desperate resistance is intended. In that case, one or a group of substantial buildings should be chosen, with thick walls, flank defence, and sheltered by other houses if possible. Its garrison may vary from 1 company to half a battalion. There should be no communication made through this keep.

No communication through it.

But a few sentries must be posted along the front line to keep a sharp look-out, and good lateral communications, as also from front to rear, provided. Sign posts and orderlies should be placed at the chief points. Some houses must be selected as guard houses; and a bomb-proof building, well sheltered, fitted up, or else constructed for a hospital.

Machine guns should be kept under cover during the enemy's preliminary cannonade; and then placed behind barricades, or to command broad streets.

The chief arrangements to be made are summarised in order of importance, viz. :—(1) clearing the field of fire; (2) concerning the shooting line, the supports, and the guns; (3) placing obstacles, and beginning (4) communications; lastly, some retrenchment or keep.

Summary.

Attack on a Village.

It is generally difficult and costly to attack a village, and should only be attempted when the object in view will justify the loss. If the village is set on fire, this will often cause it to be evacuated, but then it has to be considered whether the attacking troops may not have to pass through the village, and their advance be stopped by the fire.

Difficult, and rarely attempted.

The best plan is to concentrate a heavy artillery fire, not only to reduce the defender's fire, but to destroy buildings, and remove obstacles. Sufficient time should be allowed for the cannonade before

* Allow time for heavy artillery fire.

Storming party with engineers.

* sending on the infantry, who should be accompanied by a few engineers with tools; but the main body of engineers must be held ready to push on when the outer line is carried, to strengthen the village against any counter attack by the reserve.

The attacking troops should try to get possession of a detached building, which will screen their advance, and enable them to keep a fire from under cover, before delivering the assault. Working parties should follow to intrench any position won, and some feints should be made at points not intended to be forced, though they may be converted into real attacks if opportunity serves.

Working parties.

German attack on Le Bourget.

* The attack by the Germans, in their war with France, on the village of Le Bourget, may be quoted as an example. It was made on three sides. The two flanking columns sent out clouds of skirmishers, who gained ground at the double and lay down. The supports and reserves followed, also in extended order and at the double. When these last threw themselves down the attackers went on, bearing to the flanks. Arriving within range they lay down and opened fire. The gaps, occasioned by the extension of the front line to the flanks, were filled up, and the flanks were prolonged by single companies always in extended order, so that the attack was concentric, becoming denser and more inclosing as it approached the enemy. The characteristic to be noticed is the "rapid change from open to close order directly the most trifling cover admitted rallying a section or a company; and, on the other hand, every advance over open ground took place in widely-extended lines, moving like ants."

Defence of a Wood.

Troops should be so posted as to be able to use their rifles freely without being seen. To hold a wood defensively we must occupy the outskirts nearest the enemy, and strive to prevent his penetrating, for if he gets in the fighting will be physically on even terms, and morally in his favour. A wood will always afford infantry a useful position and shelter, and is most important when on a flank. Disorder, loss of connection and control must be guarded against. Cavalry and artillery will be generally hampered.

* Enemy must be kept out by occupation of the outskirts.

There is risk in occupying a very extensive wood, and it is then preferable to take up a position more to the rear, but still near enough to fall upon the enemy's columns attempting to debouch.

When a forest is traversed only by certain roads it affords a defending army similar advantages to mountains, for there will be long and difficult defiles, through which an enemy must pass, and we cannot only meet him coming out, but at the same time send detachments through the wood to take him in flank and rear.

Roads through forests resemble defiles.

In holding a wood obstacles must be opposed, the defenders, well sheltered, should be able to direct a strong fire on the assailants. The front line should know where the supports are posted, and good communication should be established between these two bodies and between the supports and reserve.

In some cases a wood is traversed by roads, in others by only timber tracks. A commissioned officer

L

Marking communications. *	should be specially detailed to mark the direction to different places on trees with an axe and a paint-pot, and connecting files must be carefully placed to pass orders from the fighting line to the supports, which are out of sight of the former.
Preliminary examination.	The preliminary examination or reconnaissance should note the size, length, and breadth of the wood, its nature, whether open or thick, whether the edge is marked or indefinite, if there are underwood and straggling trees between the main wood and the open, if there are any outlying clumps or belts of trees, and their distance outside the wood; the roads and paths, houses, clearings, streams, and wet places, whether the ground is broken or smooth, and what positions are in rear or on the flanks.
Abattis at salients. *	The first thing to do is to commence at once to construct abattis at all the salients, 100 to 200 yards at each place will be sufficient, leaving spaces between them of about 500 yards, through which offensive attacks may be made. Cross-cut saws work fastest. The smallest trees will be felled, but the largest left standing for cover, and to save labour. The rough abattis should be about 12 yards wide. The salients are usually on the spurs of hills, and when the enemy sees the abattis he will probable try to get round the flanks, so that the abattis at each salient should be prolonged into the wood on both sides so as to take the assailants in flank (*vide* Plate VIII.).
Rifle-pits. *	Rifle-pits should be dug to shelter the men on the flanks. Other obstacles may be placed in front, such as will not screen from fire. Detached clumps within 500 or 600 yards should be cut down or

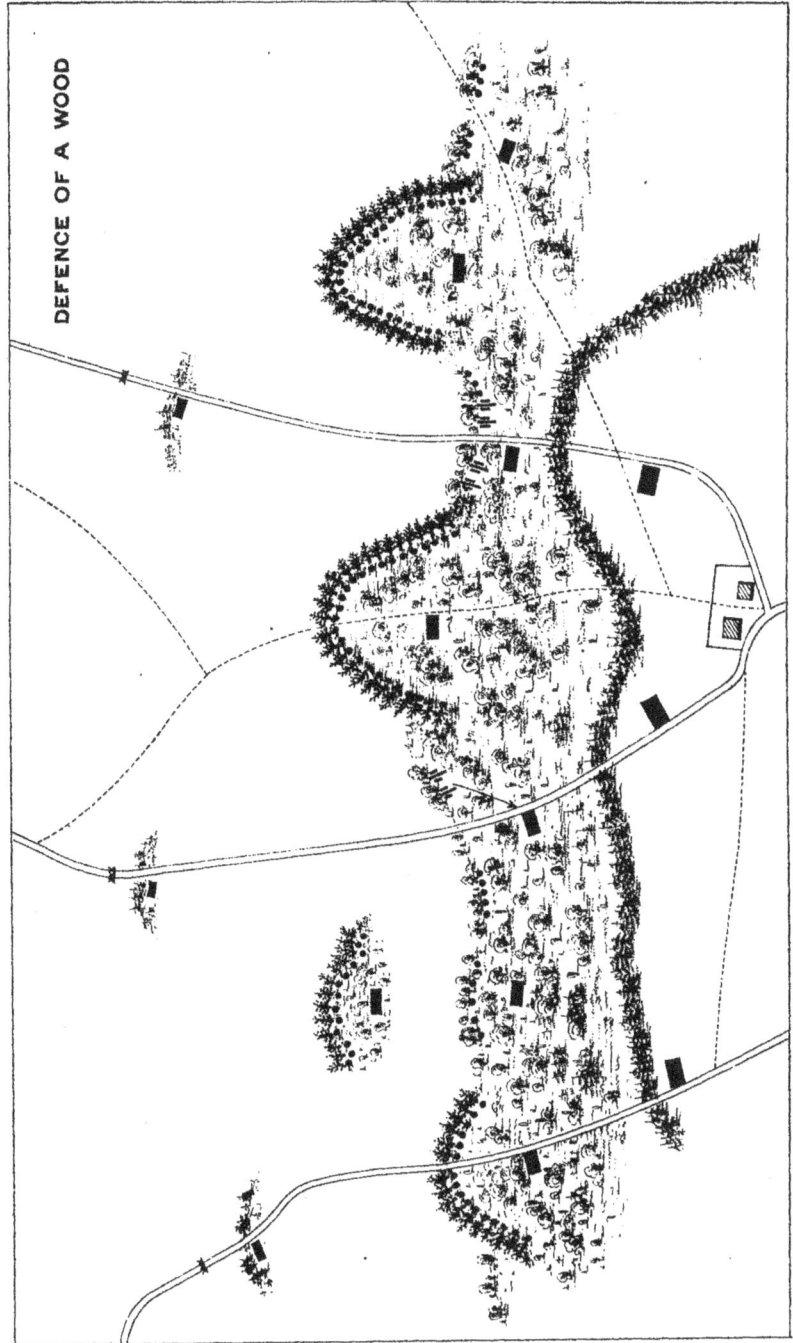

Plate VIII

occupied, otherwise the enemy will make use of them. As a rule, all cover should be cleared within that distance, but if the edge of the wood is very irregular we must decide how much to hold and how much to give up. _{* Clear field of fire.}

Unless by special order it is best not to break up or block roads at the entrance of a wood, but rather do so at a farm, or other point some 700 or 800 yards in front, where a small post might be established to cover the barricade. _{* Where roads should be blocked.}

As to guns, very few should be in the wood, when it is in front of, and flanked from the general line of battle. But if the wood forms part of that line they must not be on the roads, offering a mark, but places for them should be specially prepared, a little to one side of the roads, and some distance apart to avoid splinters. If there is time, each gun should have two or more such epaulments prepared beforehand, near enough for it to be moved from one to another by hand. But guns must have a ready means of retreat by the road. They are best outside the wood on the flanks. _{Position of guns.} _{Epaulments.}

In the American civil war, 1861-1864, there was much fighting in woods. Both armies threw up long lines of breast-works, some a simple trench and parapet, other parts felled trees laid lengthways, with small epaulments for the field guns near the paths. These works remained long after the war in the Chickahominy Swamp near Richmond, and in Georgia. _{* Breast-works in American civil war.}

The chief point in the defence is to hold the edge strongly, and to take steps to ensure prompt action of the supports and reserves, in case the front line is forced.

Supports. * The line of supports may be brought nearer to the front than in other situations. They may be of less strength, and posted in smaller bodies near the most vulnerable places. The reserve, also, must not be far away. The commanding officer should watch events from the front and send back his orders.

Second line of defence. * A second line of defence should be provided at some clearing or open space, or along a watercourse or stream, and it is advisable to make some lines of abattis parallel to the line of retreat, which will take the enemy in flank and prevent him extending, should he penetrate.

Position of reserve. * There is often a spot where all roads converge, in the woods; it may be a farm or small village. This might be placed in a state of defence, and the reserve posted near.

Telling off the garrison. * The force to hold a wood may be estimated at from 1 to 2 men per pace of the boundary. They should be told off into different sections by tactical units, and the greater part of the men placed in the fighting line.

The supports and reserves may be considered safe from fire if they cannot see the open through the trunks of trees. But if the timber is not close they should be in trenches or under other cover.

When cavalry can be employed. The employment of cavalry is restricted. They may have a space of favourable ground on a flank or in rear, from which they may come round and take the assailants in flank. In a few instances small parties may be concealed in the wood if it admits of their action.

When possible, guns should be posted so as to bring a heavy fire on a wood in front, without endangering the defenders retiring, and the latter should have their line of retreat clearly pointed out, so that they can fall back by detachments, while the enemy's troops, having attacked and gained the wood with loss, will probably be in disorder and unwilling to quit cover, especially in face of artillery fire. *Heavy fire of artillery on wood.*

If the wood be set on fire to any great extent we should evacuate it, and take up a position in rear.

Attack of a Wood.

A large wood is injurious to an army on the offensive. But small woods may be very useful, masking troops, concealing reserves, and sometimes enabling cavalry to act opportunely on the flanks of an attack.

Troops attacking a wood are in ignorance of what force there is within, how it is disposed, and how it may be sheltered. It is a dangerous and difficult operation. Tactical formations cannot be carried out, therefore an attack on a wood strongly held would be only justifiable when the object in view cannot be attained otherwise. *Dangerous operation.*

We must learn what we can of the interior from maps and the inhabitants, and reconnoitre well the front and flanks, endeavouring to know how the wood is held, and where the enemy has his infantry and guns, and where are his weak points. A false attack may make him show his position. * *Obtain all possible information.*

The defenders will probably have imperfect communication, and the supports and reserves less under control than in the open, whereas the assailants will have complete communication and supervision. It is, therefore, best to make two or three distinct though simultaneous attacks instead of one.

Two or three attacks at once.

As an attack made in the open to be successful generally presupposes superiority of force, much more is this necessary when the enemy is sheltered behind trees, walls, shelter trenches, &c.

Artillery must prepare the way with deliberation, all the guns available being brought into position, and divided into as many divisions as there are attacks. The fire should be rapid; those guns supporting all attacks may support each other for a time in succession, firing at the points selected to be forced. They may advance to within 1,500 yards, as probably there will not be many guns in the wood.

Artillery attack.

Under this fire the infantry will make the actual attack, the supports moving close up to the fighting line, and they will try to force the breach and effect a lodgment. If the edge of the wood is gained a sufficiently strong body must be collected before advancing further, and then the line must move carefully and steadily, with supports on the flanks, ready to meet any counter attack by the enemy.

Infantry.

Before issuing from a wood troops must be re-formed.

Re-form before advancing further.

Defence of a Defile.

A defile is any ground over which troops must necessarily pass with a very narrow front in proportion to their strength, so that the extent of front limits the number of combatants, and enables a weak force to contend successfully against one much

Meaning of a defile.

stronger. But a defile may be turned, or possibly avoided altogether, by the enemy discovering a mountain path, or crossing a stream by a neighbouring ford, the existence of which was unknown to the defenders holding a bridge. Hence the latter should look out well to anticipate these movements and be ready to take the enemy in flank while he is searching for another point of passage.

Although the term is usually applied to a mountain pass, or a road traversing a forest, yet it also includes railway cuttings, hollow roads, streets, causeways, embankments over swampy ground, and all bridges.

The road itself should not be cut up, nor obstacles difficult of removal be constructed, unless it is known that we shall not require the road ourselves, and it is determined to deny the use of it to the enemy. *As a rule, the road must not be cut.*

The most effectual defence is to be made by occupying a position on our own side of the defile, extending on both sides, so that every gun and rifle can be brought to bear upon the head of the enemy's column as he debouches. *Defensive position on our own side.*

Another way is by taking up a strong position somewhere within the defile, blocking the road with obstacles, and placing the men under the best available cover. This is practicable if the flanks are inaccessible, or made tolerably secure, but not otherwise, nor should it be adopted if part of the defending army has not yet passed through. *Position within the defile.*

When this is the case the defile must be held on the enemy's side, infantry being disposed according to the ground in a chain of small posts guarding the front and flanks, and the guns commanding the chief *Position on the enemy's side.*

approaches. The rest of the army will then pass to the further side of the defile, and take up the best position to protect the gradual withdrawal of the defending troops. But a position on the enemy's side should not be occupied under other circumstances, unless there is no ground in rear suitable for defence, or it is specially desired to prevent the enemy entering the defile, or part of a retreating army has not got through. For if the defenders are driven back into the defile in their rear, the retreat may be converted into a rout, and great loss entailed.

<small>Disadvantage of this position.</small>

<small>Mountain defile.</small> * In the case of a mountain defile, traversing strong ground, when the heights on each side are difficult of access, a stand may safely be made at the entrance nearest the enemy, infantry being extended along the heights commanding the defile, to make sure of the flanks, while the road is barricaded, and the infantry must make a stubborn defence. But if the heights have no command over the road, the stand must be made further to the rear, either in the defile itself or at the exit.

Attack on a Defile.

<small>Duty of infantry attacking a mountain defile.</small> * Supposing it is required to make an attack on a defile like the last referred to, so long as the defenders hold the heights, no advance is possible, therefore, parties of infantry must precede, to ascend the heights and clear them of defenders, while the

<small>Artillery.</small> guns destroy the obstacles and any cover in the defile. This done, a party of infantry will seize the

<small>Seizure of point of passage.</small> * passage, with the bayonet if necessary. The further advance must be made cautiously, flanking parties always working along the heights, and an advanced

Fig. 56. TETE DE PONT.

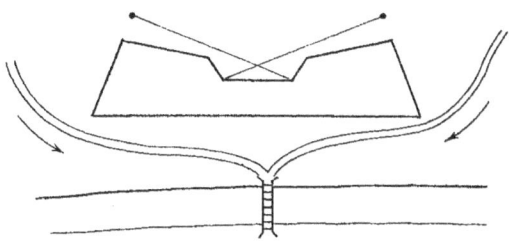

Fig. 57. SALIENT BEND.

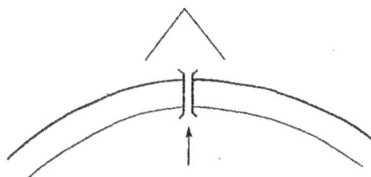

Fig. 58. RE-ENTERING BEND.

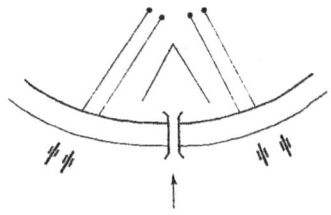

Fig. 59. STRAIGHT REACH.

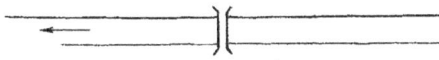

Fig. 60. USUAL POSITION OF A FORD.

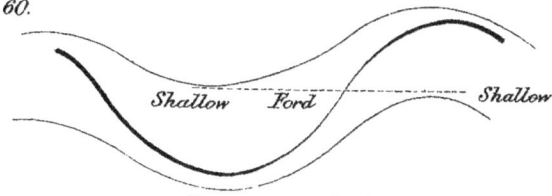

Shallow   *Ford*   *Shallow*

Deep and steep bank.

guard thrown out on the road itself, followed by some guns as near the front as is safe, ready to take position on the further side. Supports must be close up, in case further opposition is experienced.

* Caution in further advance.

Defence of a Bridge.

The same principle obtains as in the case of other defiles, and the best position is on our own side of the bridge, due precaution being taken to watch the roads on the other side. But sometimes a village is built on a river, and on both banks; then the houses, hedges, and walls must be placed in a state of defence and occupied, lest they are made use of by the enemy. If, however, we are not strong enough to do this, level the buildings on the enemy's side as far as possible, and clear the field of fire. Again, if the bridge is in rear of the village, the enemy's side must be held.

* Best defence usually on our own side.

The dispositions of both artillery and infantry should be such as to bring the greatest amount of fire on the bridge itself and the ground beyond.

* All available fire to bear on the bridge.

If a field work is made by way of a *tête-de-pont*, it must be closed, and no road leading to the bridge must pass through the work, but outside it, so that the defenders of the work may not be affected by their friends retiring, or by the enemy in pursuit.

* No road should lead through a *tête-de-pont*.

The troops retiring may assemble behind the work, and file over the bridge, while the enemy will be exposed to the fire of the *tête-de-pont* (*vide* Fig. 56).

By a comparison of Figs. 57 and 58, it is evident that a bridge situated in a bend of the river

* Best position for a bridge over a winding river.

re-entering to the enemy is much more capable of defence than one in a salient bend. But if a temporary floating bridge is constructed, a straight piece of the river will be more suitable (*vide* Fig. 59), as the current runs more easily, there are fewer eddies, and the approaches to the banks are likely to be good.

When a straight reach is desirable.

Attack on a Bridge.

The general plan of attack may be understood by reference to Fig. 38 representing a battalion advancing across a bridge in contact with the enemy (*ante*, p. 55).

Artillery cannonade previous to infantry advance.

But artillery must first cannonade his position before the infantry advances. These will move on in extended order, keeping up a continuous fire, and making the best use of cover. Should a single building be occupied defensively, a party must be told off to attack it, so as not to delay the general advance; but if the enemy is in force on our side, the houses must be shelled, and cleared of defenders, before the bridge itself can be stormed.

Defence of a Ford.

Limit of depth.

A ford should not be deeper than 3 feet for infantry, 4 feet for cavalry, and 2 feet 6 inches for artillery and waggons. The best ways to find a ford are to see where wheel tracks enter the river on one side and come out on the other; by questioning the people; by finding out what communication there is between houses on the banks facing each other; by taking the depths in a boat; by noticing

Discovery of a ford.

bends in the river where the current is strong and broken.

It is well to know that when a river flows through a bed which has not a homogeneous geological structure throughout, the current will hollow out the softer parts, the straight course will be deflected, and the current will not continue straight, but will be again deflected. This is the origin of the serpentine course of rivers. The deepest part follows the strongest current, as indicated by the thick line in Fig. 60, and is called by the Germans the "thalweg" (or valley-way). Thus a ford is generally found in the direction from one convex part to the next opposite, diagonally across the river, while the opposite concave parts will be the deepest, and the banks near them steep. [*] General direction in a winding river.

We must always defend a ford on our own side of the river. For if we defended it on the other, and were forced back, the enemy would learn where it was. As in the case of a bridge, fire must be concentrated. A ford may be made impassable by deepening it, or placing trunks of trees, stakes, harrows, or other impediments. But if we require to use the ford this should not be done. [*] Reasons for defending a ford on our own side.

Attack of a Ford.

When troops approach the river bank they must first find the ford, and the crossing will be slow and on a small front. Therefore, artillery should first commence fire, and endeavour to clear the other side. Infantry should line the bank on each side Artillery opens the attack.

<small>Infantry lines the bank.</small> * and under their fire some guns and cavalry, with infantry in support, should cross quickly, and establish themselves on the further side.

Defence of a River.

The enemy's object will be to pass over a part of his troops and establish himself on our side of the river, and under cover of these the rest of his men will cross with comparative safety. The first party over must naturally be weak, and should be easily overwhelmed. But if reinforcements succeed in following, troops should be ready to oppose them.

<small>Hold in force certain points only, but carefully watch the whole line.</small> * It is out of the question to hold a great length of river, and the enemy will assuredly adopt ruses, and feint at different points. Therefore, we should only hold in force the weakest points, e.g. bridges and fords, and dispose a series of advanced posts to watch carefully the whole line, while the main body

<small>Post reserve centrally.</small> * is posted in one or two central positions, ready to move to any point that may be threatened. Frequent

<small>Reconnoitre frequently.</small> * reconnaissances should be made, and a good system of communication established, so that counter attacks may be prepared; for a mere passive defence will avail little.

When the river is broad there will be greater difficulty in throwing a bridge and in crossing by boats. But when it is narrow, and the ground on our side is suitable for defence, it may be better to offer a weak resistance at the point of crossing, and to make the chief stand at some previously selected position across the enemy's probable line of advance.

Passage of a River.

From what has been remarked under the head of Bridges, it will be seen that when a bridge is well held and defended, an attack upon it will most likely entail loss. It is, therefore, best to choose some point of crossing in a bend of the river concave to the defenders, with firm ground on both sides, and where the bank on our side is the highest. By these means guns can be quickly brought up to direct a converging fire on the enemy's position. But they should not be brought into action until the last moment. *[Selection of point for crossing.]*

A hamlet, wood, or some other shelter should be looked for on the opposite side, which will cover the advanced party when first moved across, and enable them to hold their own till reinforced. Sometimes confluent streams and islands may be made use of in the passage. *[Shelter on the opposite bank.]*

All available boats, material for bridges, and other requisites should be collected under cover, and troops massed in rear of neighbouring heights. Detachments should also make feints to distract the enemy. *[Boats and material.]*

When all is ready a cannonade should be opened on the opposite bank, the guns being specially directed against the enemy's artillery. As soon as the other side is cleared of defenders, the covering party should be sent over as quickly as possible, and they, when once established, must strengthen their post, and defend it obstinately against any counter attack, pending the arrival of reinforcements. Besides boats, rafts may be used, and even some of the men *[Under artillery cannonade.]* *[Covering party crosses.]*

may swim across, their arms and accoutrements being carried for them.

A feint may sometimes succeed.

It may happen that a party detached to make a feint is unopposed, and an opportunity may offer to act on the enemy's flank. But if the party is small, no risk should be run of being overpowered on the other side unless they can join hands with the troops attacking at the main crossing.

Passage of the Douro, 1809.

* Perhaps the best instance of a successful passage to refer to is that of the river Douro by Sir Arthur Wellesley on the 12th May, 1809, in presence of the French army under Marshal Soult (*vide* Plate IX.).

Near Oporto the river takes a sharp bend round the Serra rock on the left bank, opposite to which was an isolated building called the Seminary. Sir Arthur Wellesley observed that his guns from this rock could sweep the opposite bank, and that its position hid the passage of barges; also, that the Seminary was a strong building, capable of holding 2 battalions, easy of access from the river, but surrounded by a high wall on the other three sides. He collected secretly his troops behind the Serra rock, placed 18 guns on the summit, and detached a brigade under General Murray to Avintas with orders to send down more boats, and if possible to cross there himself.

When Sir Arthur Wellesley heard that Murray had found boats, he sent 1 officer and 25 men over in a barge to the Seminary, and reinforcements quickly followed. 3 boat-loads crossed before the French were alarmed; they then tried to storm the Seminary, but our guns swept the left of the

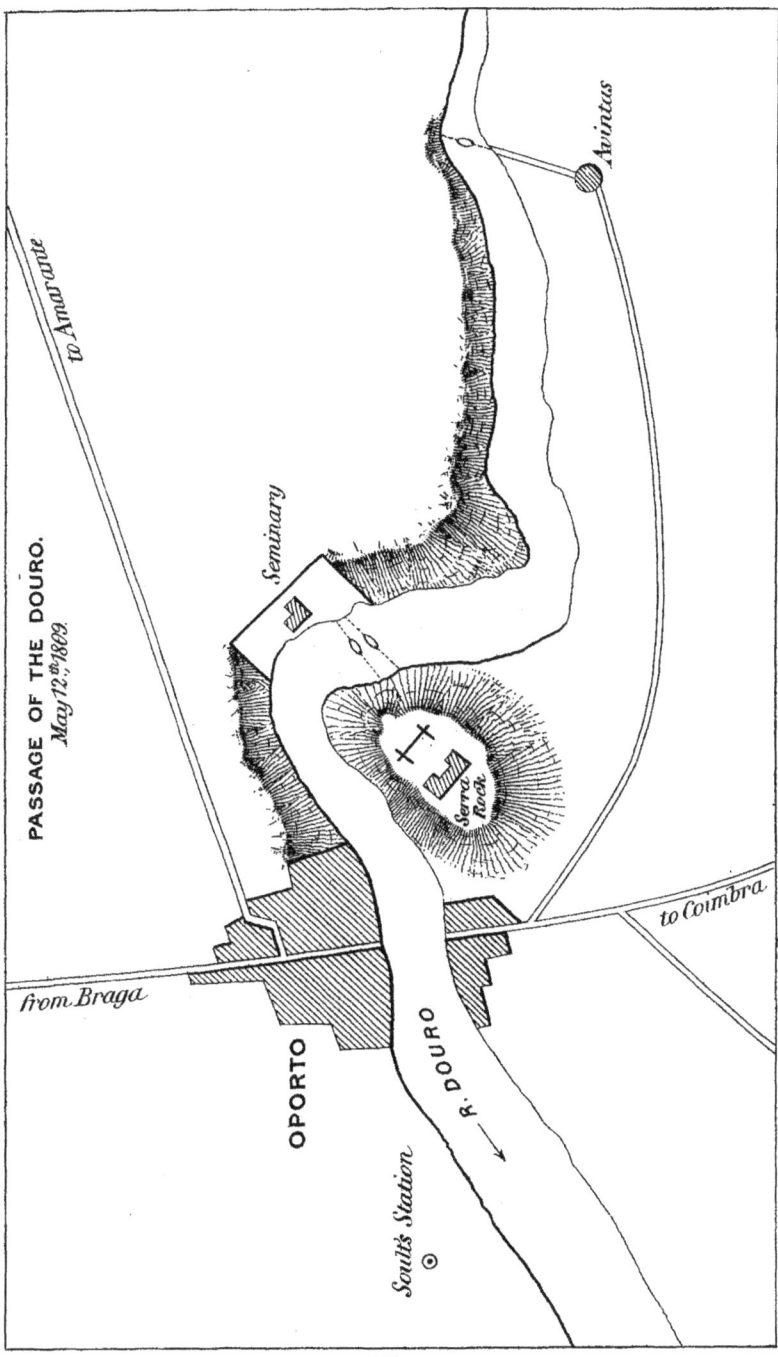

building, and confined the assault to one side, where the musketry drove them back. Meanwhile, the inhabitants brought over several great boats, by which the Guards crossed, and passing through the town took the French in rear, while Murray's brigade appeared, coming from Avintas; so the French, surprised, and fearing to be cut off, abandoned their sick and 50 guns, and retreated in great disorder towards Amarante.

The panic was so great that a squadron of the 14th dragoons cut their way through 3 battalions in a hollow road, and so complete was the surprise that Sir Arthur Wellesley sat down in Soult's quarters to the dinner prepared for that marshal.

Again, on the 7th October, 1813, Wellington crossed, with the allied army, over the river Bidassoa into France with very little loss, that river being held by Marshal Soult. Wellington manœuvred to distract Soult, and succeeded in crossing at different points, partly by a bridge left unbroken by the French, who were driven from a strong intrenched position above it, and partly by a ford near Irun, pointed out at low water by a countryman. * Passage of the Bidassoa, 1813.

The French, falsely secure from the strength of ground, neglected to watch the mouth of the river, and were surprised. * Result of a bad lookout.

Barricades.

A barricade to block up a street or road is made by cutting a ditch across it, throwing the earth into a parapet at least 7½ feet high, and facing it with * How improvised.

paving stones. But if there is little time, bags, boxes, baskets of coal, ashes, manure, sacks of corn, casks, or other solid materials from the neighbouring houses, may be formed into a parapet, while planks laid on chairs, tables, or casks may serve as the step for the defenders to stand on. Carts, carriages, wheelbarrows, and furniture may be strewed in front as obstacles (*vide* Figs. 61 and 62).

Loophole adjoining houses. * Adjoining houses should be loopholed to afford a flanking fire. To prevent a barricade in a main street being turned, any streets leading to the rear of it should be similarly blocked.

Attack of a barricade. * To attack a barricade, artillery should breach it. If guns are not available, some men should try to enter neighbouring houses, and fire from the upper rooms and roofs upon the defenders. A passage may be forced from one house to another until the rear of the barricade is reached, and a party should try to turn it by passing down another street leading behind it.

STREET BARRICADE

Fig. 61.

BARRICADE

Fig. 62.

CHAPTER IX.

OBSTACLES AND OTHER ACCESSORIES TO DEFENCE.

MORE than two hundred years ago, Sir William Temple said:—"In all sieges the hearts of men defend the walls, and not the walls the men." And this is as true now as then, with respect to holding any post or other position which it is our duty to defend. The defence of Rorke's Drift in South Africa in 1879 is a most brilliant example, where two subalterns and about 130 men successfully withstood the attacks for hours of an enemy flushed with victory, and in overwhelming numbers. The officers had presence of mind, and applied their knowledge of field fortification in the limited time and means at their disposal. * Rorke's Drift, 1879.

The object of all defensive works is to enable a weaker to resist a stronger force. When it is possible, engineers superintend troops of the line. But they may not always be available, and there are many instances where unskilled labour may be turned to account, with a little knowledge on the part of the officers, who should remember not to * Works auxiliaries to defence.

rely on the works themselves, which are but auxiliaries to the defending troops, and a passive resistance is easily overcome.

Obstacles are most efficient when placed some little distance in front of the parapet, say from 80 to 100 yards, in the case of a small work; but in front of a line of works, this distance may be greater. Then the attacking troops, detained under musketry fire, will have to overcome the obstacles, and be thrown into disorder, and then have to assault and get into the works.

Conditions * to be fulfilled by obstacles.

The conditions to be fulfilled by all obstacles, are :—They should be hidden from the enemy's distant view and fire, which is usually effected by a glacis; be defended by heavy fire; and be strong enough to form serious impediments. They should be under close musketry fire, *i.e.* within 300 yards of the defending line; but not too close, lest the defenders be disadvantageously affected by the close proximity of an attacking force. They must afford no cover to the enemy, and they cannot do this if within effective range. If possible, they should be protected from the enemy's artillery fire, so as to be intact when his infantry comes up to them; and be difficult to remove or surmount; and lastly arranged with gaps sufficiently large to allow a counter attack to be made ; *e.g.* 150 yards is enough for half a battalion to extend for attack, and for cavalry and artillery to pass.

Different * kinds in use.

The chief obstacles are abattis, entanglements, palisades, fraises, military pits, pickets, crowsfeet, gabion band trip, hurdles, chevaux-de-frise, and an inundation.

Fig 63.
ABATTIS.

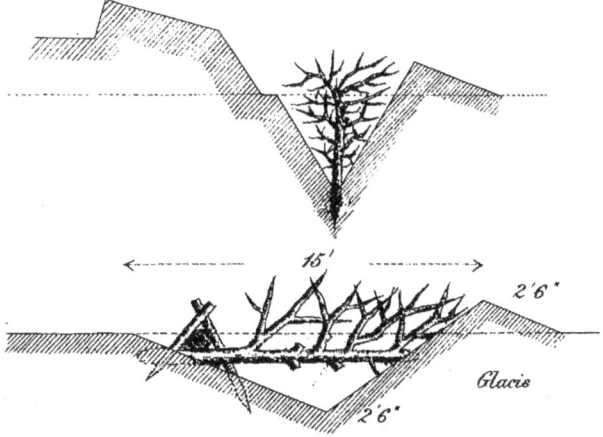

Fig 64
WIRE ENTANGLEMENT.

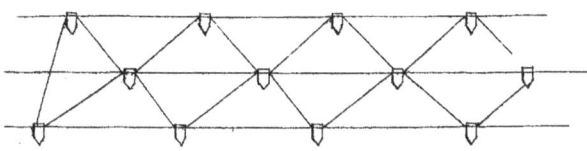

Abattis are felled trees, the trunks and stoutest limbs being cut 12 to 15 feet long, and laid as close together as possible, with the branches towards the enemy, and pointed, at least 5 feet high, with the butts buried in the ground, secured by stout stakes, or by logs laid across several butts. The smaller branches and leaves are removed, so as not to give cover. They are placed upright in the ditch, at the foot of the counterscarp; or more usually beyond the ditch in a trench, $2\frac{1}{2}$ feet deep, protected from artillery fire by an advanced glacis $2\frac{1}{2}$ feet high (*vide* Fig. 63). As an estimate of time, 6 men will cut and fix 1 tree to cover 3 yards of ground in 1 hour. And 20 men will make 2 rows, 30 yards long, in 6 hours, when the trees are small and close at hand; half the men felling, pointing the branches, and dragging the trees into position; the other half fixing them, and picketing down the butts. *[margin: Abattis.]*

The tools required are 6 felling axes, 2 hand axes, 6 bill-hooks, 2 hand saws, 2 mallets, and drag ropes.

Hard, tough wood is best. Pine is worst, being easily broken and burnt.

To fell trees men should work in gangs of 3—one with a saw, one with an axe, the third with a rope to haul the tree over on the required side, to be indicated by a nick. 3 such men in a gang will fell a moderately sized oak tree in 3 hours, and a fir tree in 10 minutes. Backwoodsmen, as in America, work much quicker with axes.

Entanglements are excellent in a wooded country, affording means of using a wood, which otherwise would give cover to an enemy. They were much *[margin: Entanglements.]*

employed in the American Civil War. They are made by half sawing the trees through, and pegging them down, so as to interlace each other.

A wire entanglement is good both against infantry and cavalry. It consists of wire stretched across stout stakes, 4 feet to 7 feet apart, arranged in rows chequerwise, the wire crossing diagonally and twisted round the heads of the stakes 1 or 2 inches above the ground. No. 14 B.W. gauge is the most convenient size—1 mile in length weighing 90 lbs. This obstacle is rapidly prepared, very portable, little injured by artillery fire, and impassable by cavalry. It is most effective when at least 10 yards deep, and concealed with brushwood and small bushes (*vide* Fig. 64).

5 men will make 50 yards of this entanglement in length, by 10 yards in depth, in 10 hours, with 900 yards of wire. They require 3 bill-hooks, 1 mallet, 2 pairs of pincers, 150 pickets.

A wood, if felled for 20 or 30 yards, forms a good obstacle alone.

Palisades. * Palisades are roughly squared timber about 6 inches square, cut through diagonally, so that the longest sides are about 8 inches, making a stout paling to guard against assault. Their usual height is 10 feet, with 3 feet buried in the ground. They are kept together by 2 ribands, one underground, the other near the top, both being on the inside, and are placed 4 inches apart, which spaces should be accurately kept, to prevent a man putting his foot through on to the riband, and so helping himself to climb over.

When the trees are too small to cut across, as in a

country with much fir, palisades are made of round timbers.

They must be covered from artillery fire; their usual position is, therefore, in the ditch, or to close the rear of a small advanced work seen into from the main line.

Fraises are simply palisades in a horizontal or inclined position, and their best place is on the counterscarp, sloping slightly down to the ditch, for then they are a formidable obstacle, and do not interfere with projectiles thrown into the ditch. Their position, when under the escarp, is not so good, as there they may assist in the assault. * Fraises.

Military pits are employed in a country affording no wood. They are either made 6 feet or $2\frac{1}{2}$ feet deep, *i.e.* either too shallow or too deep to be useful as rifle pits. The large ones are traced on the ground by forming equilateral triangles of 10 feet side, and describing circles with a radius of 3 feet, which leaves an interval of 2 feet between each 2 pits, on which the earth is heaped. The smaller pits are placed chequerwise, and the earth is spread about. It is usual to make 3 rows of each kind, and pickets are placed at the bottom with sharpened points. It will take one man a day to make a large pit, or 6 small ones. They are effective against cavalry, and break the formation of infantry (*vide* Figs. 65 and 66). * Large military pits.

* Small military pits.

Pickets are sometimes used alone, as by the Chinese in the Peiho forts. They are simply pointed stakes. * Pickets.

Crowsfeet are good against cavalry, but not much use against infantry. Each is formed of 4 iron spikes 3 inches long, joined at the heads in such a way that * Crowsfeet.

when thrown on the ground, one point is always upwards. They may also be placed in fords.

Gabion band trip. * The gabion band trip is formed of the bands of Jones's iron gabions buttoned and placed in lines 3 or 4 feet apart, connected with stout wire or rope, passed through the bridging holes, and secured at intervals to pickets. The rows are parallel and chequerwise (*vide* Fig. 67). This is a useful obstacle against cavalry, and even against infantry during a night attack.

Chevaux-de-frise. * Chevaux-de-frise are made of cast-iron pipes in lengths of 6 feet, about 5 inches in diameter, with wrought-iron spears 4ft. $7\frac{1}{2}$in. long. The pipes fit together, and the spears pack in them, so that they are easily carried in the field. Sometimes they are made of beams of wood, with spears 6ft. long, or sword blades (*vide* Fig. 68). They are easily removed, and are therefore useful to close a road. In such a position they are chained together and fastened to stakes, so as not to be dragged away. They are often used to close the entrance of a field-work, in which case one end has a wheel, and the other is fastened to a hinge on a post.

Inundations. * Inundations may sometimes be made. But if a stream runs parallel to our front, when inundated, it will be as much an obstacle to ourselves as to the enemy, preventing any advance to the attack. It is, therefore, better to inundate by means of a stream running perpendicularly to the front, for then the enemy's attack will be divided.

Dam. * The dam must be strong enough to resist the pressure of water and the enemy's attack, either by night or by distant artillery (*vide* Fig. 69).

Fig. 65.
LARGE MILITARY PITS.

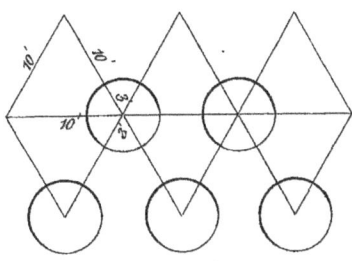

Fig. 66.
SMALL MILITARY PITS.

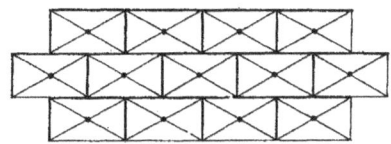

Fig. 67.
GABION BAND TRIP.

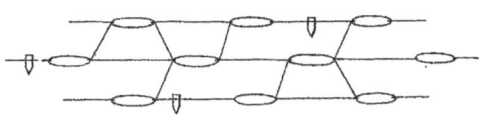

Fig. 68.
CHEVAUX DE FRIZE

Fig. 69.
DAM.

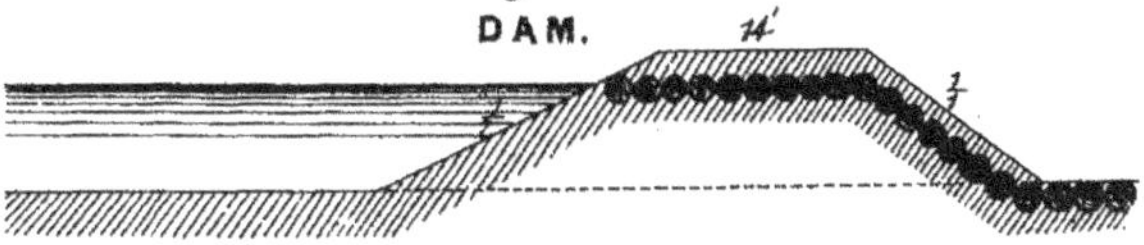

Fig. 70.
BRIDGE WITH ARCH BLOCKED UP.

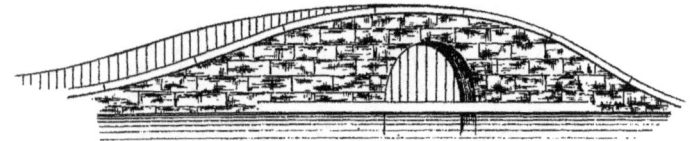

Fig. 71.
WEIR.

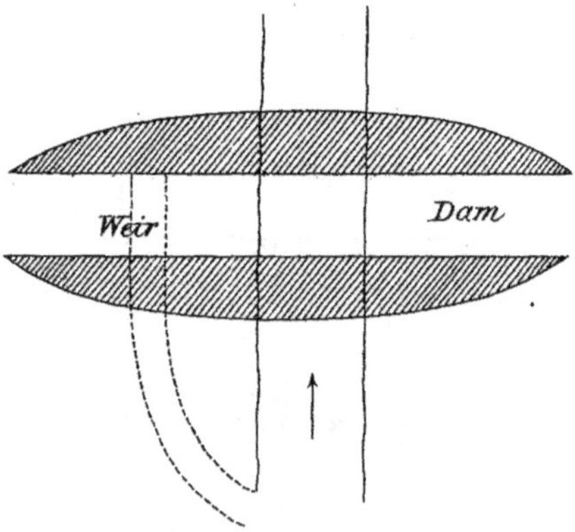

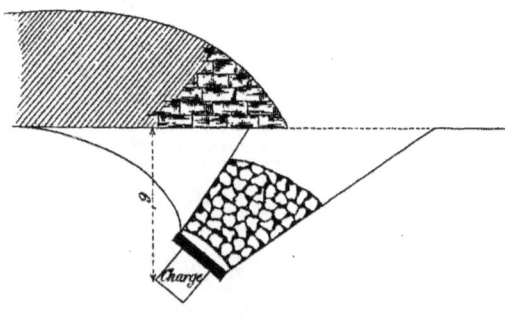

Fig. 72.
COMMON FOUGASS.

The slope next the water should be $\frac{1}{2}$; the outside slope, $\frac{1}{4}$; and the thickness at top about 14 feet at least.

By filling up an arch of a bridge with piles, an inundation may often be made, the roadway being generally raised on each side, and an overflow left under the arch (*vide* Fig. 70). Arch blocked up.

Even when the water is only 2 or 3 feet deep, a shallow inundation may be made by cutting trenches and scattering the earth.

In an inundation there must be a waste weir, cut at the level of the water, and protected from being worn away by a revetment of fascines; the outside slope and the bottom should also be revetted.

The waste weir is made at the side of the natural course of the water. The dam must extend some distance on each side of the stream (*vide* Fig. 71). The parts beyond the stream are first made, and the thickness increased gradually as it gets nearer the water. The waste weir is made at one side with a cut leading to it. The central part of the dam is filled in as quickly as possible with bags of earth, stones, &c., as many men working as close together as possible without crowding. Waste weir.

In a hilly country, an escarpment may be made, *i.e.* cutting the side of a hill or slope to make it steep. But care must be taken to have it flanked, otherwise it will shelter an enemy. It depends on the soil, for some will not stand when cut away steeply. * Escarpment.

Self-acting fougasses and torpedoes were much used by the Americans. The quantity of powder in each is about 2 or 3 lbs., placed about 6 feet below the surface. Some means must be adopted to Fougasses.

explode them as they are passed over, *e.g.* a glass tube laid on the ground, which, when trodden upon, will allow sulphuric acid to run down into a composition, and ignite the powder.

Small fougasses or land torpedoes are sometimes used to defend lines and field-works. The usual form is shown in Fig. 72.

The excavation is 6 feet deep; the earth is banked up behind and revetted to insure its acting in the proper direction. Above the charge a shield of wood is placed, and over this stones, but better still, live shells. It is fired by electricity, the wires leading to the rear. The best position is under the salients.

In the defences of Richmond in 1864, the confederates buried live shells a little under ground, and fired them by means of a sensitive fuse, which ignited when trodden upon.

Stockades. * A stockade is a line of stout timbers, planted close together, and loopholed. It not only serves as an obstacle, but gives cover. The timbers are roughly squared, and to be bullet-proof must be at least 6 inches thick of oak, 18 inches of fir. The loopholes are cut at intervals. They are planted 3 feet or 4 feet in the ground, and stand 7 feet or 8 feet above it (*vide* Fig. 73).

It is more quickly made by using round timbers without squaring them, leaving spaces at intervals for men to fire through (*vide* Fig. 74).

On the inside earth may be heaped up against the bottom, backed by logs, laid lengthways, and fastened down by stakes or planks to keep the earth up. Loopholes may also be made with sand-bags.

Fig. 73.
STOCKADE WORK.

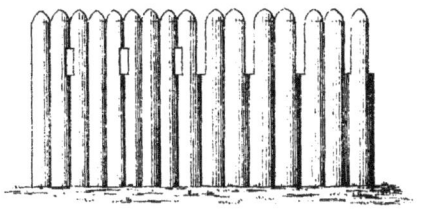

Fig. 74.
HASTY STOCKADE.

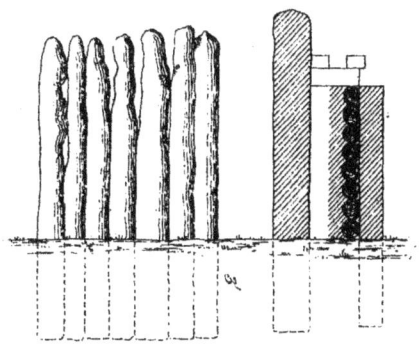

Fig. 75.
FORM OF LOOP HOLE.

Fig. 76.
LOOPHOLES ON THE TOP OF A WALL.

Stockades cannot withstand artillery fire, and it is useless to make timber proof against it, as by heaping earth against it on the outside. Also, by so doing, the use of the stockade as an obstacle is done away with, for the enemy can run up, and so climb over.

When timber is on the ground, the wood being cut, but not fashioned, a 12 feet length of stockade with double row of timbers, requiring no loopholes, can be made by 8 men in $4\frac{1}{2}$ hours.

A single row of the same length, with a ditch in front, and loopholed, took 8 men the same time.

The plan, shown in Fig. 74, required 6 men in $3\frac{3}{4}$ hours to make a length of 12 feet.

But in all these cases allowances must be made for collecting the timber.

To board up windows against musketry requires a thickness of 6 inches of oak, 10 or 12 of Scotch fir, and 18 of common fir.   * Boarding up windows.

Men should be divided into squads of 8 or 10 to collect various materials, while the rest are distributed at work.

Loopholes have a height inside of 18 inches, this being as much as a man can elevate or depress his rifle. The width depends upon the lateral range required, and so varies with the thickness of the wall. The usual width inside is 18 inches, so the rule may be followed to make loopholes 18 inches square on the inside.   * Loopholes.

Outside, the opening should be as narrow as possible, generally 3 inches. The height outside varies from 18 inches to 30 inches depending on the thickness of the wall, and the amount of elevation or depression required. Sometimes they are made wide on the outside in order to crowd as many men together as possible for firing behind a wall, as in an ordinary sized loophole a man must change his position from side to side to get lateral range, the outside opening being so narrow. But long shallow loopholes allow very little elevation or depression.

On the top of walls. * Fig. 75 shows a form which is sometimes adopted. And on the top of a wall they are often cut down, with a stone placed above them as in Fig. 76, or a log is laid above it, as in Fig. 77. When loopholes are made in buildings, they should be about 4 feet apart on the ground floor, but not so numerous in the upper rooms; about 5 feet apart will do. Some should be made with care obliquely at the angles, to fire in the direction of the capital; but the corner must not be weakened. Loopholes must be made either too high or too low for the enemy to make use of them. For the defenders the inside floor may be raised, or a step erected (*vide* Fig. 55, p. 136).

Two unskilled men will make 1 loophole in a brick wall 2 feet thick in 25 minutes, using crowbars. If the wall is 14 inches thick they can do it in 12 minutes. If the wall is of stone, the time will be longer. No detail can be given.

Of sand-bags and sods. * Sand-bags and sods are much used to make loopholes quickly on the top of a parapet or shelter trench, in shelter pits, rifle pits, &c. The sods must be cut thick and square, and placed as in Fig. 78, with a splay to the inside.

Walls. * Walls of soft stone or brick are not much injured by artillery fire, unless it is concentrated on a wall. Single shots make a hole but will not bring the

Fig. 77.
LOOPHOLED WALL.

Fig. 78.
SANDBAG LOOPHOLES.

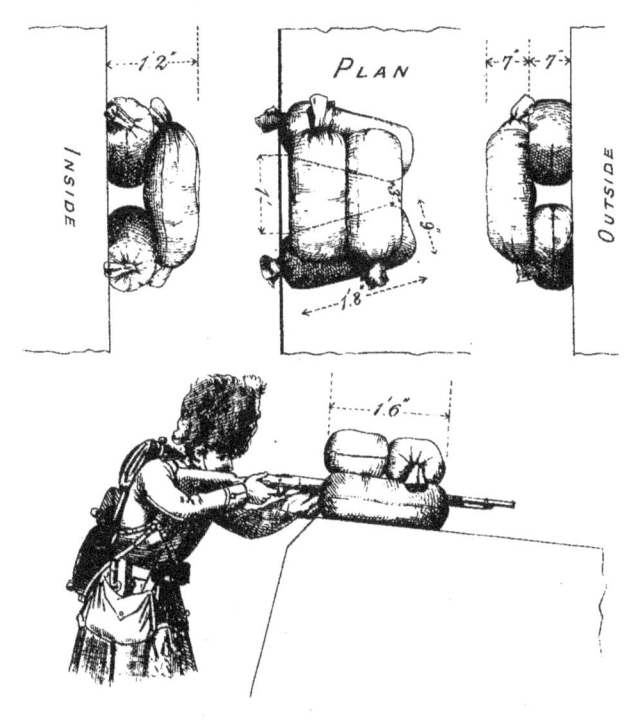

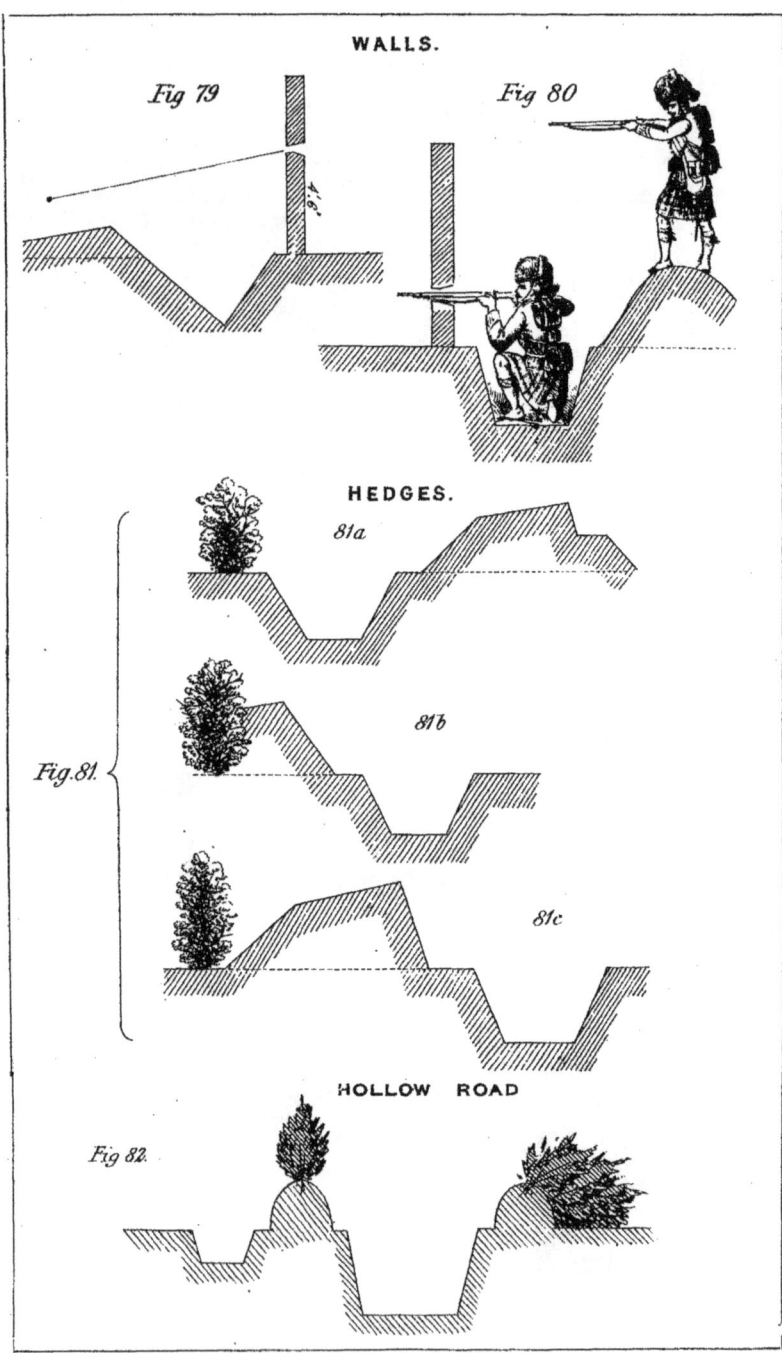

BREASTWORKS

Fig 83   Plan.

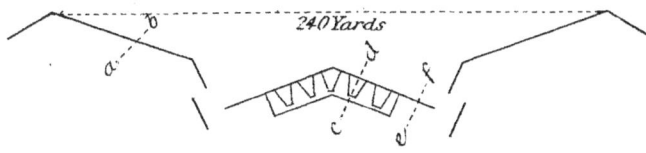

Sections

Fig 84

Fig 85.

Fig 86

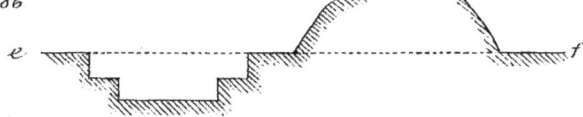

Fig 87.

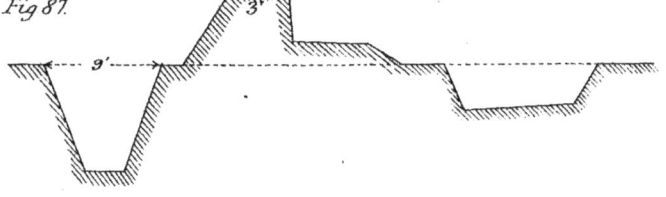

whole wall down, and soft stone does not splinter much, therefore we may safely use such walls as a screen, as an obstacle, and as protection against rifle fire, but not against artillery.

Earth should not be heaped up outside the wall unless it is very high. The best plan is to dig a ditch at the foot of the wall, and throw the earth forward into a glacis commanded by the fire from the loopholes (*vide* Fig. 79).

If the wall is 6 feet high the loopholes should be cut 4 feet 6 inches from the ground; and the ditch in front must be made, else the enemy will close with the loopholes. Or a banquette may be used, and the loopholes made higher.

When a double row of loopholes is required, the best method is to dig a trench about 4 feet deep inside, from which some men can fire through the row made low down, while the earth from the trench is formed into a mound in rear, from the top of which the others can fire (*vide* Fig. 80).

If the wall is too high for men to fire over it standing on the mound, we must erect scaffolding, but this is likely to be knocked down by artillery, so it is better to break down part of the wall.

When the height of the wall is 4 feet 6 inches, trunks of trees supported by stones may be laid on the top (*vide* Fig. 77), or loopholes may be made by placing large stones or sods on the top.

In making use of hedges it is preferable to treat them as obstacles, and to leave a hedge in front of a work, having the ordinary parapet and ditch all behind it (*vide* Fig. 81*a*).   *Hedges.*

In such a position the hedge forms a sort of

entanglement, and a serious obstacle for men to get through under fire.

A low parapet may be thrown up in rear of the hedge, the earth being got from a trench (*vide* Figs. 81 *b, c.*). This gives cover, while the hedge is still an obstacle and a screen. At Hougomont, on the field of Waterloo, the hedge was about 20 yards in front of a loopholed wall, and the French were shot down in great numbers trying to get through it.

Hougo-
mont.

Hollow
road.

In the case of a hollow road, the road should be used as a ditch; a trench being dug behind the hedge on one side, while the hedge on the other side is bent down to form a sort of abattis or entanglement (*vide* Fig. 82).

Improving
banks, &c.

In improving banks and hedges, a working party may be extended at wide intervals; for 1 man can convert 20 to 30 feet of hedge into a good breastwork in 3 hours.

Breast-
works.

Breastworks are the simplest and quickest parapets that are thrown up in the field. Figs. 83 to 92 are rough diagrams of some of their forms. Fig. 83 shows how the line is broken for flanking defence. The next three are sections of different parts. Fig. 88 shows the advantage of availing oneself of any undulation of ground, which is apparently unimportant. Fig. 92 is a form for rocky soil with 2 lines of workmen.

Labour.

It is considered that 1 man excavates 1 cubic yard per hour. The number of hours varies according to the soil. Generally 6 cubic yards is the work of a day of 8 hours.

The usual and convenient distance for workmen

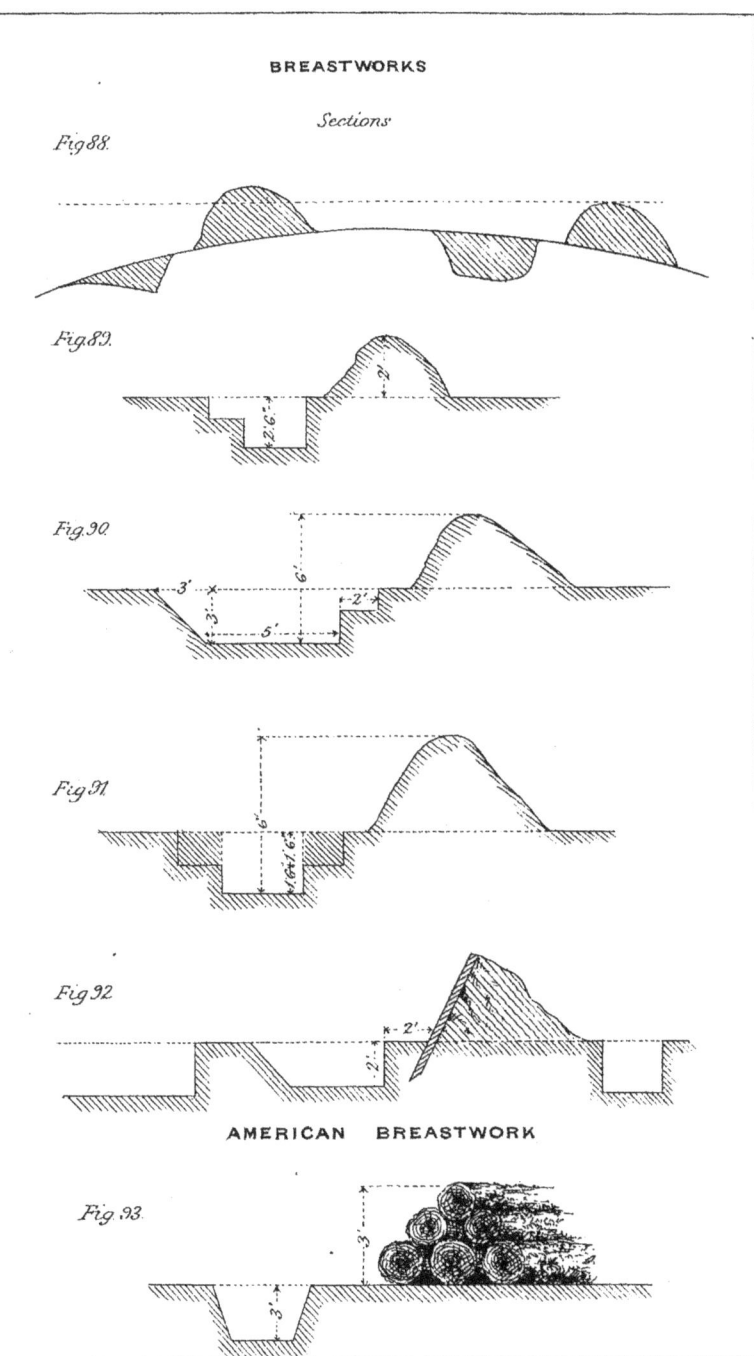

to be apart is 6 feet. But this is not invariable, and is sometimes reduced to 4 feet; but time is not gained in proportion.

Steps should always be made for easy communication over the parapet, and also in rear when there is a trench, to enable the defenders, after firing their last round, to retire, and interpose a fresh obstacle to the enemy. *Communications.

The tools and stores required for strengthening posts are embraced under the following heads, viz. :— *Tools and stores.

i. Field-exercise tools—*i.e.* shovels, picks, felling axes, bill-hooks for digging trenches, making breastworks, abattis, &c.

ii. For houses and walls—sledge-hammers, hand-borers, crowbars, saws, augurs, spike-nails.

iii. General service — sand-bags, rockets, hand-grenades, small shells.

Each man in the construction of a breastwork has 1 pick and 1 shovel.

One man can carry 100 sand-bags, the weight of which when empty is 60 lbs. Each bag holds 1 bushel.

In the American Civil War troops always intrenched themselves, and with rare exceptions the defenders repulsed the attack. The men worked in squads, varying in number according to the nature of the wood; but it was generally pine. Each squad felled 6 trees; the branches were lopped off, and the trunks laid as in Fig. 93, to form a breastwork, giving about 3 feet of cover—a trench 3 feet was dug, and the earth thrown over. The men then set to work to clear the field of fire *American breastwork.

by cutting down as many trees as possible in front. This was their intrenchment when time was short.

<small>Petersburg.</small> * But at Petersburg and Charleston, when more time was available, bomb-proof barracks were constructed, beams being placed in the ground, and covered by a strong roof of timber and earth not visible outside. Sometimes the casemate was in the parapet. Thus the men were kept well under cover from artillery fire, and always close at hand.

<small>Bomb-proof barracks.</small> * For the construction of bomb-proof barracks, a trench is dug of a depth depending on the nature of the soil, about 5 or 6 feet, and drained, its width being 12 feet. The sides are lined with timber 6 feet long, and 1 foot thickness of timber placed on top. The earth is kept up behind the timber by planks, slabs, and hurdles; and over the whole the earth is piled. To get into this steps are made at intervals, and lighting is obtained from windows between the level of the ground and the roofing timber. The men sleep with their heads to' the wall, and a little passage is left between them. In case of a heavy fire, it may be necessary to roof it over the edge, projecting beyond the steps (*vide* Figs. 94 and 95). Similar bomb-proofs were made in the last war between the Turks and Russians

<small>Plevna</small> * —*e.g.* at Plevna.

As earth will not stand at a steeper slope naturally than about $\frac{1}{1}$, *i.e.* with a base equal to its height, materials are used to support it at a higher angle.

<small>Use of revetments.</small> * These are called revetments; and are gabions, fascines, hurdles, planks, sand-bags, sods, &c.

As a rule, it is only the interior slope of a parapet

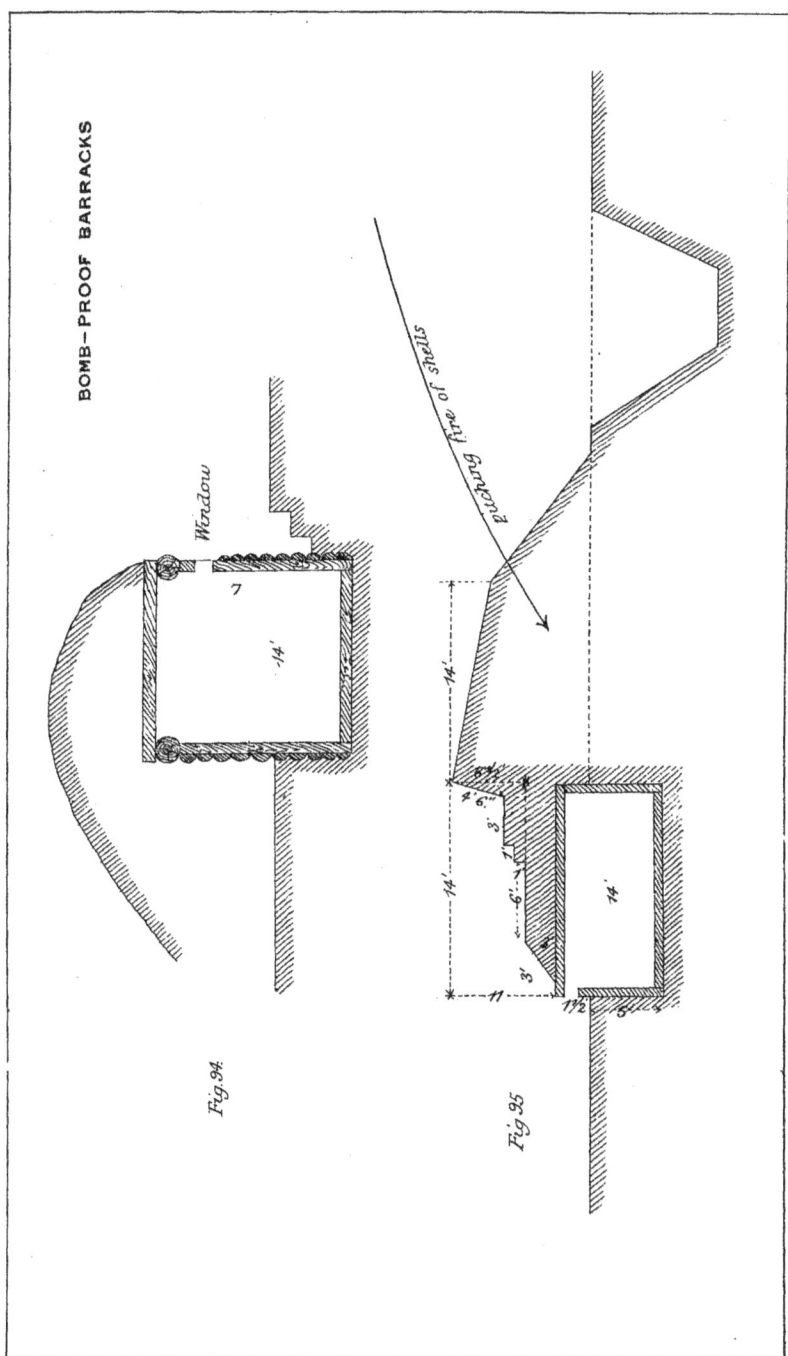

that must be revetted, so that men can stand close up to it to fire. Brushwood will make gabions, fascines, and hurdles. The Chatham calculation is that 50 men can fell an acre of wood in 7 or 8 hours, about a day's work, including packing it in waggons. Fourteen waggons are required to carry the quantity cut by them; and if it is ordinary brushwood, it will make 100 gabions, 40 fascines, and 2,500 pickets.

* Chatham calculation.

A gabion is a strong cylindrical basket without top or bottom, 2 feet in diameter, and 3 feet high. To make one, describe a circle with a piece of cord and two pegs 11 inches apart as radius, and divide the circumference with the cord into 6 parts. Place 6 pickets with the thick ends downwards at these points, and other 6 pickets in the intermediate parts of the circle, with the thin ends down. This will make the basket of uniform size. Flexible twigs or rods are then interwoven with the upright pickets, commencing with 3 rods at the bottom, and weaving each in succession outside of 2 pickets and inside one. As they are expended, others are added, and the basket-work or web is continued to the height of 2 feet 9 inches. It is then sewn in 3 or 4 places from top to bottom with withies called gads, or spun yarn, to prevent it coming off the pickets. The ends of the pickets are cut off about $1\frac{1}{2}$ inches beyond the web, and pointed. The best wood for the web, and particularly for the gads, is willow or hazel. A gabion when made stands 3 feet high in revetment, weighs 36 to 40 lb., and is made by 3 men in 2 hours. It is the most durable revetment, resisting the shock of the discharge of guns in embrasures

* Gabions.

better than fascines. They stand in position by themselves, and can be formed into a revetment by ordinary labourers.

3 practised men will make 4 gabions in a day of 8 hours. 2 gabions and an 18-foot fascine make nearly the same amount of revetment.

The division of labour in making a gabion is for 1 man to weave, 1 man to keep the pickets upright, and 1 man to prepare gads, by treading on one end, and twisting them well till they are supple.

Figs. 96 and 97 show the method of making, and Fig. 98 the way they are built up in revetment, a layer of fascines forming the foundation, and another layer between two courses.

When the parapet is only 4 feet 6 inches high, there will be 1 foot 6 inches height remaining above a single gabion. It is better to revet this with sods or sand-bags than with fascines, for a single shot will knock away a whole fascine.

Jones's iron gabions.

Jones's gabions (*vide* Fig. 99) require each 10 pickets and 10 bands of galvanised iron, each about 6 feet long. 2 men put a gabion together in from 10 to 15 minutes, as follows:—One band is fastened and laid on the ground, and 4 pickets are placed (2 inside, and 2 outside), dividing the circle equally. Then the rest of the pickets are fixed alternately inside and outside. The bands are placed one by one over the pickets, inside and outside alternately, the slots and buttons being all kept to one side. Each band must remain half way down the pickets until the next one is put on; this will ensure the pickets continuing in their places. The dimensions of a gabion are 3 feet by 2 feet. These gabions are

Fig. 96.

GABIONS

Fig. 97.

Fig. 98
GABION REVETMENT.

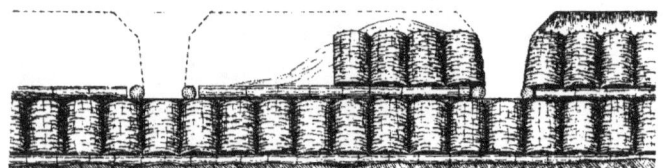

Fig. 99
JONES' CABIONS

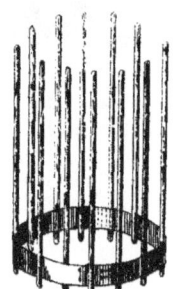 

Fig. 100
CARRYING CABIONS

MAKING FASCINES

Fig 101.

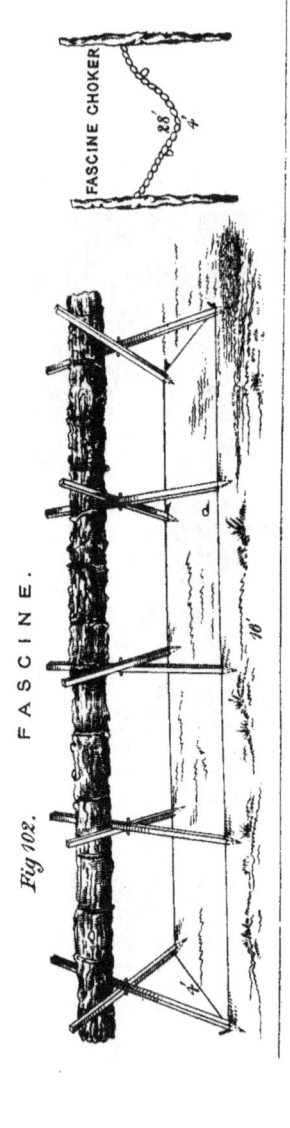

Fig 102. FASCINE.

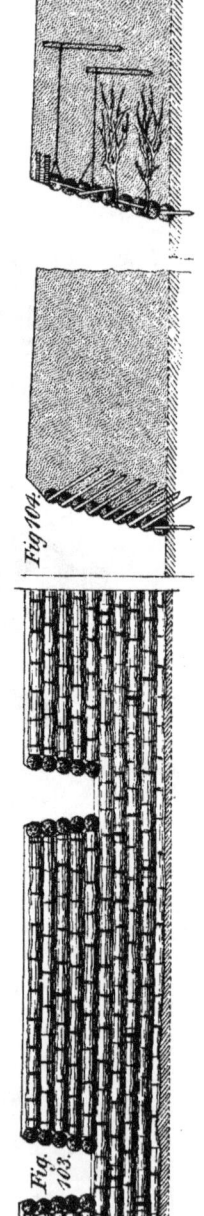

FASCINE RIVETMENTS

placed along the line of the 2nd parallel, a berm being left of 3 feet, in a siege. The slots must be placed on the front side of the parapet, to lessen the danger of splinters.

Fig. 100 shows the manner of carrying gabions to the line traced out for the parallel. Here the men are extended 4 feet apart. The picks are laid with the iron between the gabions in the ground, the shovels parallel to the tape. When half full the gabions should be slightly tilted outwards to a slope of $\frac{1}{4}$. Cover should be obtained in from 10 to 20 minutes. * Mode of carrying gabions.

A fascine, or military faggot (*vide* Fig. 101), is made 18 feet long, and 9 inches in diameter. It is sawed into convenient lengths of 9 feet, 6 feet, and even 3 feet, the last being to strengthen the space between 2 gabions. The best way to make a fascine is to trace with string and pegs a rectangle on the ground 16 feet by 4 feet, and to set up at equal distances 3 to 5 trestles, according to the strength of the brushwood. Care must be taken to have the trestles cross at the same height; this can be done by stretching a string between the end trestles. The brushwood is laid on, the smallest inside, the longest and straightest outside, and projecting at each end 18 inches. It is then compressed to a diameter of 9 inches by a choker, *i.e.* a 4-foot chain with a wooden lever at each end; and on the chain are 2 rings marking a length of 28 inches. The brushwood is bound with tough, much twisted gads, or wire at 6 inches beyond the end trestles, and at intermediate distances of 15 inches, before the choker is relaxed: and the * Fascines.

ends of the fascine are sawed off square beyond the extreme trestles (*vide* Fig. 102).

A squad of 5 men will make a fascine in 1 hour. If the brushwood is dry it will weigh about 150 lbs., if thick and green 200 lbs.

Picketing fascines. * To form a revetment they are laid horizontally in the direction of the parapet, the first row having a groove cut for it, and being fastened down with stakes; and each fascine above is similarly fastened by a picket driven into the parapet. These pickets are from $3\frac{1}{2}$ feet to 4 feet long, and there should be 1 per yard or 7 to each fascine. They are driven obliquely downwards through the fascine at an angle of 45° with the slope, and two of them are driven vertically to fasten each row to the next underneath (*vide* Figs. 103, 104). Fascines support earth at $\frac{4}{1}$, but are heavy, and require trained men to build them up. But though inferior to gabions they are more manageable, and are often used for the cheeks of embrasures. If brushwood is cut ready, 4 ordinary men will make 3 fascines in 1 day.

Hurdles. * Hurdles are quickly made of rough stakes driven into the ground, with brushwood intertwined. Common hurdles can be fastened by anchoring pickets, but are not so strong.

Planks. * Planks are easily procurable from houses, and make a tolerable revetment.

Sods. * Sods only support earth at $\frac{3}{1}$: a revetment requires much labour and time, but they are easily procurable. The usual dimensions of sods are 1 foot 6 inches long, by 1 foot wide by $4\frac{1}{2}$ inches thick. They are generally laid grass downwards, headers and stretchers, like bricks in a wall. They can be economised by laying them all stretchers, or cutting them diagonally across, and placing the front inside; but this is not so strong. Heather, straw, rushes, and clay, moistened and well rammed, may be used; also stones, but there is risk from splinters. Barrels are not good, because of the space left where they bulge out.

24 sand-bags will revet 10 feet of surface. Their dimensions are—empty, 2 feet 8 inches by 1 foot 2 inches; filled, 2 feet by 9 inches. They are portable, and often carried on board ship, when a battery can be quickly formed on shore. When filled they are bullet-proof, and very useful for loopholes. If used in embrasures they ought to be covered with raw hides as protection from the discharge of the gun. Sand-bags are perishable unless tarred, which adds to their weight. They support earth at $\frac{4}{1}$. *Sand-bags.*

CHAPTER X.

WORKING PARTIES, ESCALADE, HASTY DEMOLITIONS, ETC.

Working Parties.

Tracing a field work. * To trace a field-work the angles are marked by stout pickets, and joined by tapes or cords.

Profiles are set up by tracing perpendiculars to the line of parapet where required, and setting up laths according to the shape (*vide* Fig. 105).

Profiling laths. * In each face there should be 2 sets of profiles; or on a long face 3 or more. To make an oblique profile, first set up 2 sets on one face forming the angle, and another set on the line bisecting the angle, dressed with those on the face. When the salient angle is 60° the profile will be twice as wide as one on the face.

Two ways of forming working parties. * There are two ways of forming working parties:—

(1) The ordinary method, when the greatest amount of work is required from each man irrespective of time.

(2) When time is the great object, and it matters not how many men are employed to do the work in a fixed period.

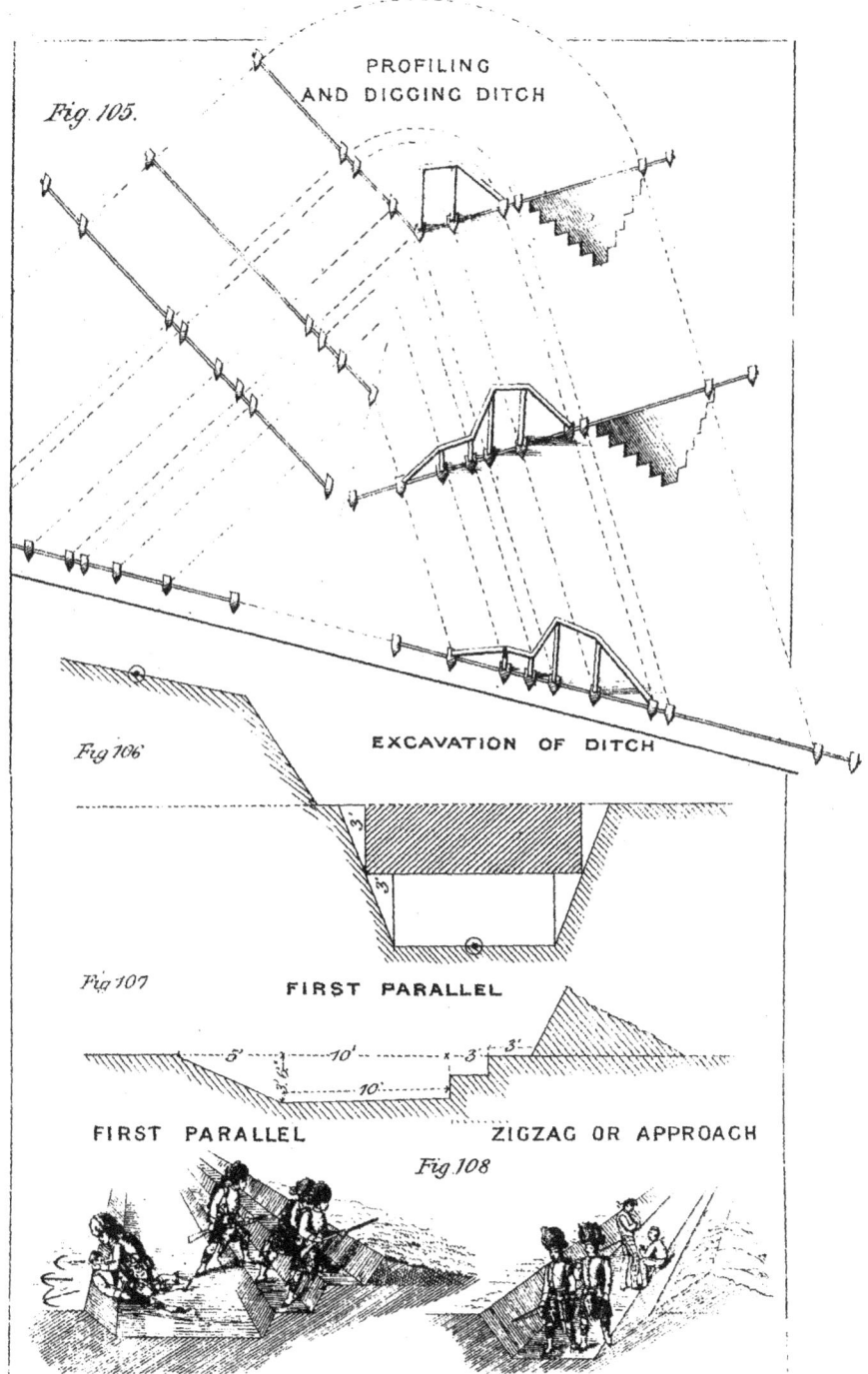

In the first case, the men must not be crowded, especially at night. The distance of 6 feet apart is fixed to let them use the shovel well; and the simplest distribution is to make all the front rank diggers, $\frac{2}{3}$ the rear rank shovellers, and the remaining $\frac{1}{3}$ rammers, which has this advantage, that as the shovellers and rammers together are equal to the diggers, they may relieve the latter.

The men are extended along the work at the rate of a file for every six feet; and if we consider the men divided into squads of 6, each will consist of 3 diggers, 2 shovellers, and 1 rammer.

It is usual to dig the ditch in steps with a base and height in proportion to the slopes; and it is divided into 2 terraces, or if very deep into 3. The men dig vertically down one step, then leaving the width of the step they dig the rest (*vide* Fig. 106). The steps thus serve as spaces to throw the earth upon, and afterwards may easily be cut away and the slopes smoothed. * Object of steps in the ditch.

The diggers throw the earth as far as they can, and the shovellers pass it on to the part furthest from the ditch; the rammers walk about over the parapet, consolidating the mass by treading as well as ramming.

Men can throw earth 6 feet vertically or 12 feet horizontally; and each of these distances is a "relay." The number required is found with sufficient accuracy by measuring the distance between the middle point of the top of the parapet and the middle point of the bottom of the ditch; *e.g.* suppose the ditch to be 12 feet deep, and the parapet 12 feet thick, there will be 24 feet from centre to centre = 4 * Relays.

relays; and besides the diggers who are 1 relay, we must have 3 relays of shovellers to pass the earth along.

This calculation is only requisite in a work of some size, and the ordinary distribution serves in most cases.

If wheelbarrows chance to be procured, much labour and time would be saved.

A certain number of the men must be told off for revetting.

Parallels. * An example of the second case, when time presses, is the construction of the 2nd parallel, each man filling 2 gabions, only 4 feet, and another in shelter trenches, when men are extended at 2 paces = 5 feet.

The 1st parallel in a siege is always executed by infantry, as also are the zigzags and approaches (*vide* Figs. 107, 108). The working parties are in 3 reliefs of 8 hours each. The ground is reconnoitred, and the direction of the parallel, whether curved or straight, is predetermined, and traced by tapes. The pickets are white, to be seen at night, and each tape of 50 yards is coiled diagonally (as for string to fly a kite). 25 workmen and 1 sapper are detailed to each tape, each man having 2 yards of trench. The sapper superintends with a 6-foot measuring rod. A berm of 3 feet is left, which is afterwards cut into a step. The trench when completed is 15 feet wide at top and 10 feet at bottom.

Extension of working parties. * Suppose the working parties drawn up in rear of the right flank of the parallel. The command is given—" By successive companies, right turn, left wheel," when 1 pace in rear of the tape, " Left wheel —right form in extended order."

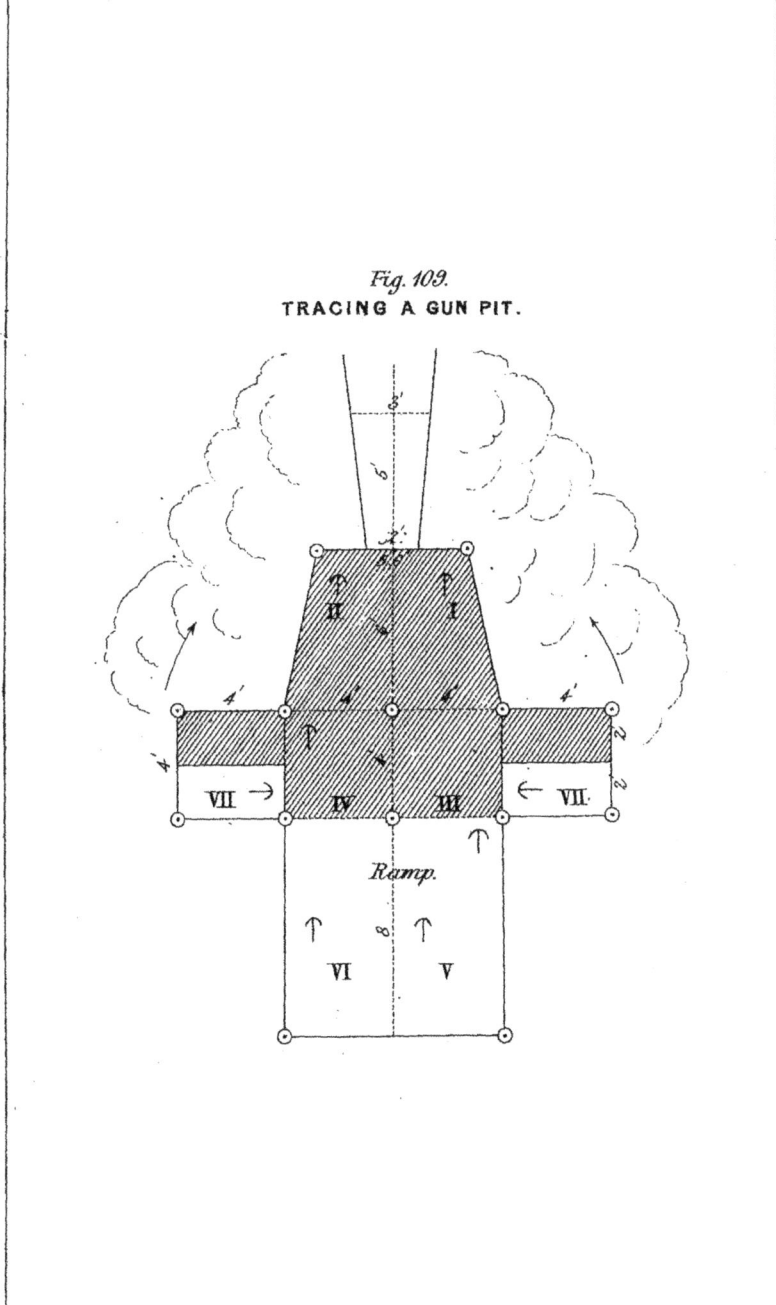

Fig. 109.
TRACING A GUN PIT.

The sapper, taking the measuring-rod hand over hand, marks off the 6 feet length of task for each man. Every man places the point of his pick in the ground on the left of his task in line with the tape, and the shovel with the blade touching the pick along and inside the tape, and lies down till the extension is completed. Word is then quietly passed to begin.

The proper way to excavate, and to avoid accidents at night, is to pick a hole on the left of the task, and gradually in grooves from front to rear, working to the right. Thus the men will keep always the proper interval. Arms are grounded or piled in rear; tunics taken off, and laid folded in regularity before commencing work.

To trace a gun pit (*vide* Fig. 109), the first thing to do is to mark off the line of fire, and the general shape of the embrasure, which is 2 feet wide at the neck, and 3 feet wide at a distance of 5 feet to the front.   * Gun pit.

A party of 7 men are required, distributed as in the figure—

Nos. I. II. III. IV. excavate 2 feet deep, and throw the earth to the front and round the shoulders.

Nos. V. and VI. make a ramp from 0 depth in rear to 2 feet. No. VII. leaves a berm of 2 feet at each shoulder, cut into a step, if time permits, for the gunners to sit upon. When the earth is 1 foot 6 inches high begin the embrasure, for the depth of the pit 2 feet + 1 foot 6 inches = 3 feet 6 inches, the height at which a field gun fires. It may be revetted with sand-bags, &c., subsequently. The time required is 1 hour.

The hasty field-work represented in Fig. 110

Hasty field-work. * was made by 4 companies at the North Front, Gibraltar. Each face and each flank was 40 yards in length; and as $\dfrac{40 \times 3'}{5'} = 24$ men, that number was extended along each at 2 paces. The salient angle was 150°; the shoulder angles 120°. One company was told off to each face or flank. The remaining men constructed a trench for the reserve, with 2 faces parallel to the faces of the work, so traced that their prolongation cut the inner ends of the flanks.

The work was traced with 3 tapes, two of them being 3 feet apart marking the berm, the third 10 feet in rear, to mark the rear cutting line of the trench, which was 2 feet deep. When the trench was nearly excavated pickets were placed at intervals along the crestline, and a tape stretched 2 feet high. The earth, brought close to the tape, was levelled and made as firm as its lightness would allow to form a base for the gabions, which were carefully placed, touching each other, with the slots to the front, and filled. The extremities of the flanks, and the tops of the gabions were crowned with sand-bags, for which purpose the gabion pickets were hammered level with a shovel.

Rifle pits. * Rifle pits are much used in sieges, in advanced positions, and for protection of a covering party, as when men are extended in front of a line being intrenched, and in the absence of a natural cover they first make shelter pits in a few minutes (*vide* p. 19, Fig. 18), and if they have more time, these are improved to hold 2 men, or a group of men (*vide* Fig. 111).

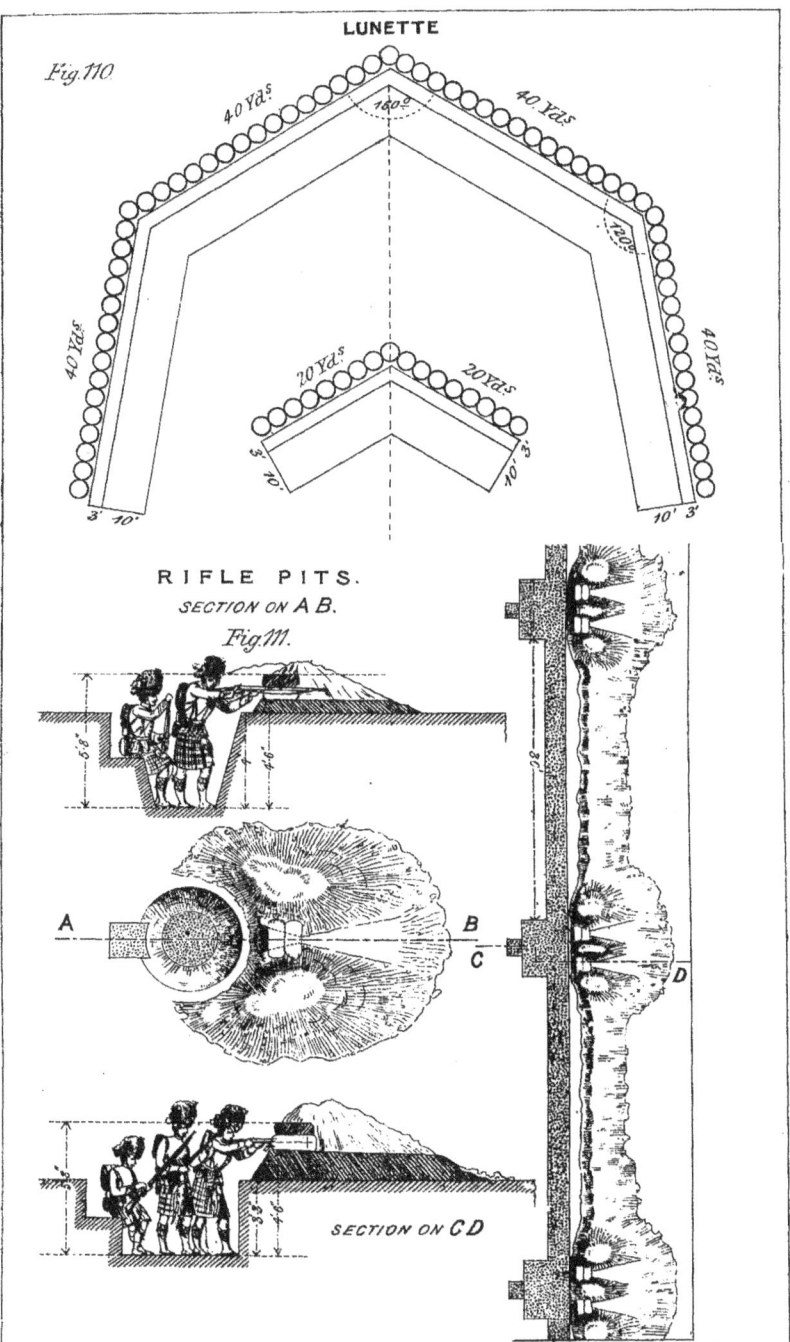

For each rifle pit 1 pick, 1 shovel, and 4 sand-bags are required, and the time required is $1\frac{1}{4}$ hours. Sometimes a trench is dug 3 or 4 feet deep connecting two or three together.

Escalading and Assaults.

Previous to the assault, whether by escalade or otherwise, of any fortified place, the exact obstacles to be overcome ought to be discovered, as far as possible, by personal observation. They will generally consist of escarps, counterscarps, palisades, abattis, and quickset hedges on the berm. The depth of the ditch gives the height of the scarps, and is most important in escalade, determining beforehand the length of the ladders. These should be long enough to allow an inclination of $\frac{5}{1}$, and 3 feet above the top of the wall. With greater inclination there is risk of the ladders breaking down with the men's weight. 28 feet is the greatest height considered possible to escalade; 30 feet is beyond it. Previous reconnaissance.

Unless the enemy be very contemptible, to be successful, an escalade must be more or less of a surprise. For this reason the men should be thoroughly instructed in what they have to do, and taught the drill beforehand, thereby avoiding confusion, and no words of command will be necessary before discovery. Limit of escalade.

Midnight is the best time when the work is to be destroyed; but an hour and a half before daylight, when it is intended to hold it, since more time is required to destroy a work than to effect a lodgment in it. It is desirable to make several Time.

attacks; and false attacks should be strong enough to be converted into real ones, if opportunity offers. The officers must know individually where they ought to lead their parties so as to concentrate. By neglecting this precaution our assault failed on Bergen-of-Zoom, when the French defeated the parties in detail. A party should be told off to open the gates.

Division of assaulting column. * An assaulting column consists of four parts; the covering party, the party of engineers and pioneers with tools to overcome obstacles, the ladder party, and the storming party, comprising the advance of the main party, and the reserve. The ladder party should not mount.

Covering party. * i. The covering party advances in extended order in two divisions, sufficiently apart to allow the ladder party to pass between them. They advance silently to the edge of the ditch, and lie down, and must not fire till the assault is discovered, but then they keep an incessant fire on the defenders, and remain in their places to cover retreat in case of failure. Firing must cease when their comrades have mounted the ladders.

Engineers and pioneers. * ii. The duties of the pioneers and engineers with tools depend upon the obstructions to be overcome. They must remove palisades and fraises below the counterscarp before the ladders can be placed. They follow the covering party, or else move in rear of the first division of ladders to clear away any obstructions in the ditch or on the berm. But if none of these exist, they should be in rear of the advance of the storming party to destroy any obstruction within the work.

They should be accompanied by a party of artillery with nails and hammers to spike any guns flanking the ditch.

This party will be about 20 to 60 men, provided with crowbars, hammers, axes, picks and shovels, bags of powder, or boxes of gun-cotton, to blow in gates and stockades.

iii. The ladder party consists of as many divisions as there are walls to be ascended or descended, the principle being to have a continuous road by means of ladders. * Ladder party.

Some narrow ditches may be filled up, as at Sevastopol, with bags of wool, or a portable bridge may be used such as the French had at the Mamelon. This is preferable to using ladders.

iv. The advance of the storming party is commanded by an infantry officer, who should be accompanied by an engineer officer understanding the interior of works. This party follows about 30 paces in rear, with the arms slung butt downwards. The drill is to get up and down the ladders with unfixed bayonets, to fix them on the berm, and when a party of 6 or more have collected, to advance together. If there is no berm the men must scramble up as best they can, assisting each other by the grip, called "the butcher's hook." * Storming party.

An engineer is told off to hold the top of the ladder to the exterior slope by a sap-fork, and the men pass down by the right of the ladder. 30 paces more to the rear the main body follows, under the officer, who has charge of the whole operation.

A separate reserve is held ready about 100 paces in rear of the main body, or conveniently posted to

Working party. * assist any one column, when there are several, as may be necessary.

A working party follows to destroy the work, or to effect a lodgment. They do this work under protection of the storming party, and may or may not be armed, at the option of the general.

It is well to supply the ladder party with revolvers and cutlasses to give them confidence. Sailors are most serviceable. But their arms must not interfere with their special duties.

Sometimes bamboo ladders are prepared on the spot, as in China. They are much lighter than those used in drill.

The latter are 24 feet long; each weighs 133 lbs., and requires 6 men to carry it; an 18-foot ladder weighs 100 lbs., and requires 4 men.

Old sand-bags should be tied round the points of the ladders to muffle them, and also to allow them to slip down the counterscarp quietly, and be placed against the opposite wall without noise.

Four companies are practised at the drill as follows :—

Escalading drill. * No. 1, told off as covering party, on the command "The escalade will commence," advances, extending half a company right and left, and lies down along the counterscarp, keeping clear of the front of the ladders.

No. 4, the storming party, retires to a position under cover.

Nos. 2 and 3, told off into sections of 6 files, form the ladder divisions. The sections are sized from right to left. After breaking into column of sections the N. C. officers commanding in rear tell off from the front.

Fig.112.
ESCALADE.

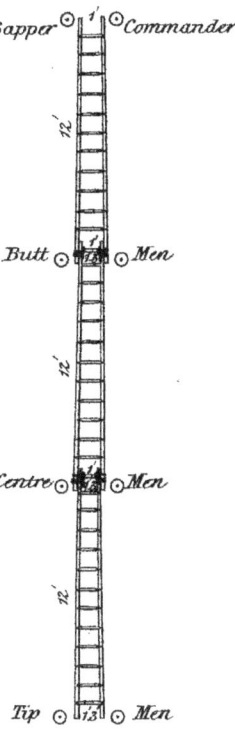

Spare Men

Arms.

tip, which they carry on their shoulders, the spare men assisting.

At "*Turn Over*," the ladders are turned by the centre and tipmen, and laid lightly against the escarp, not leaning to either side.

At the word "*Form on the Berm*," the men mount the ladders, ascend, carrying the rifles at the "short trail"; they fix bayonets on the berm, and form there, or make their way to the superior slope. The officers lead, and when provided with sword-knots these are twisted round the wrist.

Hasty Demolitions.

The explosive materials available in the field are: gunpowder, guncotton, and in mining districts occasionally dynamite. * Materials.

Guncotton and dynamite may be considered twice as powerful as gunpowder when not tamped, and four times when tamped.

Dynamite can be safely and easily used to destroy a bridge by laying a train of cartridges of No. 1 dynamite along the crown, or to remove an obstruction, or blow up a wall, by placing the train along the base, and firing it by a piece of Bickford's ordinary fuse, compressed with pincers into a smaller cartridge called a primer, placed in contact. The velocity of the explosion and pressure per square foot is very great, and no tamping is required; but if there is time and opportunity to bore holes, dynamite may be economised, as a single * Dynamite.

cartridge in a bore hole will do as much work as half a dozen laid outside.[1]

Destruction of palisades with powder. * To destroy palisades with axes is tedious. 30 to 40 lbs. of powder laid at the foot of the palisades will blow in about 4 or 5, making a breach 5 feet wide. Or a bag containing 20 lbs. of powder may be placed in a hole 18 inches deep outside the palisade, and the hole filled up. But this takes time, and cannot be done by a man under fire. Another plan is to connect two bags of powder, each containing 16 to 20 lbs., with powder hose, attached to a board resting against the palisades, and 4 feet apart. Each bag of powder should be covered with 4 sand-bags. The force of the explosion will then be greater, and none of it will be lost through the intervals of the palisades.

Destruction of stockade. * In the demolition of stockades it has been found that when there was a single row of timbers 1 foot square, well spiked together, 30 lbs. of powder made a hole large enough for a man to creep through, but 100 lbs. made a breach 15 feet wide. In both cases a few sand-bags were laid outside. A sand-bag holds 40 or 50 lbs., so the charge should be in 2 or 3. But leather bags are better for the purpose.

New Zealand pahs. * The pahs in New Zealand were made of double rows of stockade, consisting of stout oak timbers, 4 feet being between the rows. 200 lbs. of powder

---

[1] "At Chatham, in 1882, 2 lbs. of dynamite were exploded by Sir F. Abel against an iron plate ½ inch thick, with the result that a ragged hole ¼ yard long, and several inches in width, was torn through the metal. When 8 lbs. of gunpowder was used instead of ¼ that quantity of dynamite no effect was produced, the plate being not so much as bulged."—*Extract from the "Standard," November 1, 1883.*

employed against these created a breach of 9 feet 6 inches in the outer, and 5 feet 6 inches in the inner row.

At Ghuznee a gate was blown in with 300 lbs., and in China 160 lbs. destroyed a gate in one of the forts.   * Gate at Ghuznee.

But guncotton is more usually employed, being very portable. It is carried in discs or slabs, which may be safely cut smaller with a sharp knife if pressed together between boards, to keep them from coming to pieces. They are issued on service to cavalry, who can use them to advantage in raids.   * Guncotton.

Guncotton is uninflammable when steeped with water. But when a detonator, in immediate contact with a slab, explodes, the slab and all touching it, though wet, explode. A little dry guncotton, however, is necessary as a primer.

Dry guncotton should be used for small charges and in blast-holes. The ordinary Bickford's fuse usually burns at the rate of 3 feet a minute, and is best lighted with a vesuvian. But if kept long in store it burns more quickly, and should therefore be tested. It may also be lighted with portfire or a percussion-cap. The instantaneous Bickford's fuse burns at the rate of 30 yards a second, and should be used with a piece of the ordinary fuse, to allow time for the man lighting it to get away. It is distinguished by having orange-coloured worsted twisted round it. This fuse is quite waterproof.   * Bickford's fuse.

To ensure an explosion the primer should be quite dry, the charge closely touching the object to be destroyed, and the Bickford's fuse well home in the tube of the detonator and secured.

The leather bags, or tarred sandbags, containing a

o

charge of powder, should be placed within other tarred bags, to guard against a premature explosion from a spark. The fuse should be fixed in the middle of the bag.

Blowing in a gate. * A good way to blow in a gate is to hang 50 lbs. of guncotton against it by means of a sharp pickaxe, or by attaching it to the end of a ladder, which a man crosses to ignite it (*vide* Fig. 113). The charge may also be laid on the ground.

Destruction of trees. * 5 or 6 ounces of guncotton is sufficient to blow down a tree 1 foot in diameter. For larger trees the number of ounces varies with the square of the diameter. Thus, a tree 2 feet in diameter will require $2^2 \times 6 = 24$ ounces, placed in one or more holes, bored horizontally at the required height. Or the discs may be hung round the tree like a necklace, but then 8 times the quantity is required. The tree can be made to fall in any direction by attaching a rope to the branches and hauling.

Destruction of stockade. * A stockade can be destroyed by 3 lbs. per foot run of guncotton slabs placed in contact on a deal board laid against it. Similarly, upright supports of wooden bridges, iron telegraph-poles, arches, and walls can be destroyed with guncotton.

Destruction of iron girder bridge. * To destroy an iron girder bridge the charge should be placed beneath the supports.

Destruction of houses. * From 40 to 50 lbs. of powder, or less of guncotton, will destroy houses if placed under stairs, or between two thick walls. That quantity of powder destroyed stone houses in Montreal during a fire.

Destruction of a railway. * The best way to destroy a railway temporarily is to remove a couple of rails at a curve when possible. But if time presses, a 2 lb. slab of guncotton, cut

In each section Nos. 1 and 2 are spare men, No. 3 tipmen, Nos. 4 and 5 centre men, No. 6 buttmen. The numbers are proved and arms piled.

The sections are turned to the left in column, the spare men remain, the rest march to the park and fetch the ladders, which are laid in three pieces, butts to the front, in front of the sections.

At the commands, "*Shoulder Ladders*" and "*Ground Ladders,*" there must be no noise. The commanders of sections alone speak, in a low tone. The front rank are on the left of the ladders, the rear rank on the right. They move off in succession from the right into position in front of the spare men.

At "*Commanders to the Front,*" commanders of sections double out 12 paces, turn about, and cover on the ladders.

At "*Join Ladders,*" No. 6 runs out with the base section of the ladder, places the butt at the N. C. officer's foot, and runs back to raise the other end 18 inches above the ground by the rear rung.

At "*Two*" the centre man runs out with the middle section and joins it.

At "*Three,*" the tipman runs out and joins the upper portion, all being assisted by the sapper allotted to each ladder.

The ladders are in lengths of 12 feet, 1 foot clear at the tip, 1 foot 3 inches at the base.

The positions of the sapper, commander, and men are shown in Fig. 112, annexed.

At "*Lashings,*" each man takes a lashing, holding the point in his right hand, and turning inwards looses the lashing.

At "*Prepare to Lash*" "*Lash*," he stoops on the inner knee, kneeling on the side of the ladder, pulls the eye-splice towards the tip of the ladder, takes one round over all on the flat rung, one round under all, and makes fast with two half-hitches.

The commanders inspect the ladders.

On "*File on Arms*," the sections countermarch outwards and fall in by the arms, which are unpiled, and all the sections are turned to the head of the column.

Arms are slung from the "advance," the front rank on the left shoulder, the rear rank on the right, having the inner shoulders available for carrying the ladders.

Divisions. * The odd numbers of sections 1, 3, 5, form the right division, the even numbers the left. The ladders are shouldered and the rear rank men are 1 pace behind the front rank men to avoid clashing of rifles. The right division advances, and the ladders are rested just beyond the counterscarp. The sections form in file behind the ladders.

When the ladders are lowered into the ditch the base will not be more than 18 inches from the base of the wall.

At the command "*Form in the bottom of the Ditch*," the commander is the first to descend, followed by front and rear rank men alternately, all by the right of the ladder.

At "*Cross the Ditch*," the buttmen lay hold of the 2nd rung, and raising it 6 inches, move the ladder very steadily across the ditch, placing the butt against the foot of the escarp; the centre men receive the centre as it descends, the tipmen the tip, which they carry on their shoulders, the spare men assisting.

At "*Turn Over*," the ladders are turned by the centre and tipmen, and laid lightly against the escarp, not leaning to either side.

At the word "*Form on the Berm*," the men mount the ladders, ascend, carrying the rifles at the "short trail"; they fix bayonets on the berm, and form there, or make their way to the superior slope. The officers lead, and when provided with sword-knots these are twisted round the wrist.

Hasty Demolitions.

The explosive materials available in the field are: gunpowder, guncotton, and in mining districts occasionally dynamite. * Materials.

Guncotton and dynamite may be considered twice as powerful as gunpowder when not tamped, and four times when tamped.

Dynamite can be safely and easily used to destroy a bridge by laying a train of cartridges of No. 1 dynamite along the crown, or to remove an obstruction, or blow up a wall, by placing the train along the base, and firing it by a piece of Bickford's ordinary fuse, compressed with pincers into a smaller cartridge called a primer, placed in contact. The velocity of the explosion and pressure per square foot is very great, and no tamping is required; but if there is time and opportunity to bore holes, dynamite may be economised, as a single * Dynamite.

cartridge in a bore hole will do as much work as half a dozen laid outside.[1]

Destruction of palisades with powder. * To destroy palisades with axes is tedious. 30 to 40 lbs. of powder laid at the foot of the palisades will blow in about 4 or 5, making a breach 5 feet wide. Or a bag containing 20 lbs. of powder may be placed in a hole 18 inches deep outside the palisade, and the hole filled up. But this takes time, and cannot be done by a man under fire. Another plan is to connect two bags of powder, each containing 16 to 20 lbs., with powder hose, attached to a board resting against the palisades, and 4 feet apart. Each bag of powder should be covered with 4 sand-bags. The force of the explosion will then be greater, and none of it will be lost through the intervals of the palisades.

Destruction of stockade. * In the demolition of stockades it has been found that when there was a single row of timbers 1 foot square, well spiked together, 30 lbs. of powder made a hole large enough for a man to creep through, but 100 lbs. made a breach 15 feet wide. In both cases a few sand-bags were laid outside. A sand-bag holds 40 or 50 lbs., so the charge should be in 2 or 3. But leather bags are better for the purpose.

New Zealand pahs. * The pahs in New Zealand were made of double rows of stockade, consisting of stout oak timbers, 4 feet being between the rows. 200 lbs. of powder

---

[1] "At Chatham, in 1882, 2 lbs. of dynamite were exploded by Sir F. Abel against an iron plate $\frac{1}{2}$ inch thick, with the result that a ragged hole $\frac{1}{2}$ yard long, and several inches in width, was torn through the metal. When 8 lbs. of gunpowder was used instead of $\frac{1}{4}$ that quantity of dynamite no effect was produced, the plate being not so much as bulged."—*Extract from the "Standard," November* 1, 1883.

employed against these created a breach of 9 feet 6 inches in the outer, and 5 feet 6 inches in the inner row.

At Ghuznee a gate was blown in with 300 lbs., and in China 160 lbs. destroyed a gate in one of the forts. *Gate at Ghuznee.

But guncotton is more usually employed, being very portable. It is carried in discs or slabs, which may be safely cut smaller with a sharp knife if pressed together between boards, to keep them from coming to pieces. They are issued on service to cavalry, who can use them to advantage in raids. *Guncotton.

Guncotton is uninflammable when steeped with water. But when a detonator, in immediate contact with a slab, explodes, the slab and all touching it, though wet, explode. A little dry guncotton, however, is necessary as a primer.

Dry guncotton should be used for small charges and in blast-holes. The ordinary Bickford's fuse usually burns at the rate of 3 feet a minute, and is best lighted with a vesuvian. But if kept long in store it burns more quickly, and should therefore be tested. It may also be lighted with portfire or a percussion-cap. The instantaneous Bickford's fuse burns at the rate of 30 yards a second, and should be used with a piece of the ordinary fuse, to allow time for the man lighting it to get away. It is distinguished by having orange-coloured worsted twisted round it. This fuse is quite waterproof. *Bickford's fuse.

To ensure an explosion the primer should be quite dry, the charge closely touching the object to be destroyed, and the Bickford's fuse well home in the tube of the detonator and secured.

The leather bags, or tarred sandbags, containing a

o

charge of powder, should be placed within other tarred bags, to guard against a premature explosion from a spark. The fuse should be fixed in the middle of the bag.

Blowing in a gate. * A good way to blow in a gate is to hang 50 lbs. of guncotton against it by means of a sharp pickaxe, or by attaching it to the end of a ladder, which a man crosses to ignite it (*vide* Fig. 113). The charge may also be laid on the ground.

Destruction of trees. * 5 or 6 ounces of guncotton is sufficient to blow down a tree 1 foot in diameter. For larger trees the number of ounces varies with the square of the diameter. Thus, a tree 2 feet in diameter will require $2^2 \times 6 = 24$ ounces, placed in one or more holes, bored horizontally at the required height. Or the discs may be hung round the tree like a necklace, but then 8 times the quantity is required. The tree can be made to fall in any direction by attaching a rope to the branches and hauling.

Destruction of stockade. * A stockade can be destroyed by 3 lbs. per foot run of guncotton slabs placed in contact on a deal board laid against it. Similarly, upright supports of wooden bridges, iron telegraph-poles, arches, and walls can be destroyed with guncotton.

Destruction of iron girder bridge. * To destroy an iron girder bridge the charge should be placed beneath the supports.

Destruction of houses. * From 40 to 50 lbs. of powder, or less of guncotton, will destroy houses if placed under stairs, or between two thick walls. That quantity of powder destroyed stone houses in Montreal during a fire.

Destruction of a railway. * The best way to destroy a railway temporarily is to remove a couple of rails at a curve when possible. But if time presses, a 2 lb. slab of guncotton, cut

LADDER BRIDGE TO CROSS A DITCH OR BLOW IN A GATE.

Fig. 113.

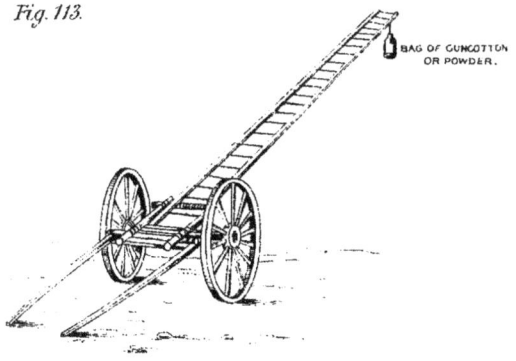

Fig. 114.
FINDING DISCHARGE OF SMALL STREAM.

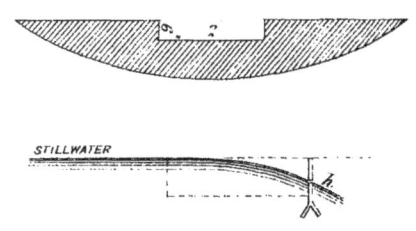

Fig. 115.
BRIDGE OF CASKS.

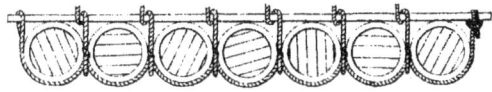

Fig. 116.
TRESTLES.

into three parts, and placed close against a rail between two sleepers, will destroy any rail.

It must be remembered that no railway or line of telegraph is to be injured except by express order. _{* Express order required before injuring railway or telegraph.}

In further damaging a railway, the rolling stock should be removed; rails taken up and thrown into a deep river or pond; or heated over a fire made of the sleepers; they will then bend by their weight; but if a pick be inserted at each end, and be borne down upon in opposite directions, the rails can be so twisted as to be rendered useless until rolled again. The station should be burnt; stacks of fuel set fire to or removed; water-tanks and water-supply rendered useless; tunnels blown in, &c.

For the destruction of a line of telegraphs the wires should be scraped bright and clean, and bound together with fine wire; instruments and batteries may be removed; insulators broken; wires cut; poles cut down or destroyed with guncotton; or they may be pulled down by a rope thrown over the wires, and cutting the stays with pincers. _{* Destruction of telegraphs.}

When the wires run underground we must search for the test boxes, or cut a trench some 4 feet deep across the probable direction of the line, dig out the pipes, destroy or bend them; pull out and cut to pieces the wires.

A field gun may be destroyed by $\frac{1}{2}$ lb. guncotton being exploded outside near the muzzle. _{* Destruction of a field gun.}

Directions for Packing Mules with the Otago Pack Saddle.

There are 2 mules to each company, with pack saddles complete. The 1st mule carries 3 bell-tents, _{* Load on each mule.}

o 2

3 peg-bags, 6 poles, 12 storm-pegs, 6 storm or guy-ropes, 2 picks with handles, 1 spade, and 1 shovel.

The 2nd mule carries 2 bell-tents, 2 peg-bags, 4 poles, 8 storm-pegs, 4 guy-ropes, 2 picks, 1 spade, 1 shovel, 1 felling-axe, 1 hand-axe, 1 bill-hook, 6 camp-kettles and lids, 6 leathern buckets.

Mode of packing. * To pack a mule—Take the straps off the saddles, and place them on the ground, and proceed as follows:—No. 1 tent, with peg-bag on top, and storm lines between tent and bag. Outside these place the two poles of No. 1 tent, and one pole of No. 3 tent, taking care that the iron and pointed parts of the poles are to the rear of the package. Outside all place the spade, pick (handle taken out), and storm-pegs. Strap the package, and place it on the off side of the mule on the ground.

No. 2 tent, and one pole of No. 3 tent, is packed in exactly the same way. Each package is then lifted at the same time. The one on the off side is hooked on first, and supported until the other on the near side is hooked. Then place No. 3 tent in the centre, with peg-bag and storm-ropes, and pass the leather surcingle over all. The buckle of the surcingle to be on the near side, in the centre of the package.

The 2nd mule is packed just the same, except that the felling-axe is placed on the off-side package, and the hand-axe and bill-hook on the near side. The 6 camp-kettles and 6 leathern buckets ride in the centre in place of the 3rd tent on the 1st mule.

Unpacking. * To unpack—The mules are drawn up in line at 8 paces interval. Undo the surcingle, double it, and place it on the mule's neck. Take off No. 3 tent and bag, and place them 4 paces on the proper left of the mules—take off the packages, and place them on the ground close to the mules. Unstrap the packages; take off the straps, and put them on the saddle; move the mules to the front, and the tents are ready for distribution. The poles of No. 3 tent are returned to that tent.

With this equipment there are 4 tents for each company—3 for the men and 1 for the officers; and No. 1 company carries a tent for the quarter guard; No. 2 carries one for the field officers; No. 3 one for the staff; No. 4 one for the rear guard.

CHAPTER XI.

CROSSING RIVERS, AND BRIDGING.

Fords. * The methods of judging approximately the position of a ford have already been noticed under that head (*vide* Fig. 60 p. 155). A gravelly bottom is best, for sand or mud get stirred up, and the depth is increased. A ford should be marked by long pickets driven into the river bed above and below, and connected by a strong rope.

A bridge constructed across a river will dam it up more or less and alter the current, the velocity of which must be observed on this account, as also to provide for the security of a bridge, making inundations, calculations for supplying them with water, and for the supply of camps.

Velocity of stream. The velocity of a stream is greatest in a straight part of its course near the surface, and near the centre. It is least near the banks and near the bottom. A rough estimate, true for surface velocities up to about 4 miles an hour, is that the mean velocity $= \frac{4}{5}$ the surface velocity, and the velocity at bottom $= \frac{3}{5}$ surface velocity.

To find the surface velocity we may use a patent log. But a simple way is to float some body, *e.g.* a piece of stick, which nearly sinks and is not affected by wind, and see the time it takes to float past two marks on each bank, or, if the river is too wide to see this, then we must anchor two boats.

A good plan is to float a light rod weighted at one end, with its tip above water, and note the distances floated in a certain number of seconds, then $\frac{7}{10}$ the mean number of feet a second gives the number of miles an hour.

A sluggish stream runs 1 mile an hour, an ordinary stream 2 miles, a rapid one 3 miles, very rapid 4 to 5 miles, and a torrent 6 miles an hour. These are surface velocities.

A convenient way to ascertain the discharge of a small stream is to dam it across with sods and insert a board with a notch in it, without overflowing the sides of the dam (*vide* Fig. 114). _* To find discharge of a small stream.

While the water is flowing through this measure the height of the still water above the bottom of the notch, stooping down to look through the notch. Then if h be that height in inches the discharge per minute $= 10 \sqrt{10h^3}$ in gallons for each foot the notch is wide; *i.e.* double this amount in the instance shown in Fig. 114, the notch being 2 feet wide.

The allowance of water in a standing camp is 5 gallons a day per man, and 8 to 12 per horse for all purposes. * Allowance of water.

Different kinds of bridges are used on different rivers * Different kinds of bridges.

Pontoons, boats, and casks make good floating bridges, when the depth is not less than 2 feet and the banks are not too high, otherwise it would be difficult to get on the bridge. If the boats are good the velocity of the current does not matter much, but casks offer more obstruction.

Rafts can be used to cross the same sort of river, but not in a swift current. A trestle bridge can be conveniently made in a river not exceeding 7 feet in depth, and the current not more than 3 miles an hour, with a good bottom. Piles are suitable for a depth even of 10 or 12 feet, if the bottom will do for driving them, and is not rocky nor very soft.

Suspension bridges, or rope, or wire, are difficult to make, but are sometimes convenient for crossing deep chasms where there is no support from the bottom. Spar bridges are available for openings up to 40 or 45 feet, with sound, high banks. A bridge can be made over a shallow slow river with firm bottom with carriages, also with fascines and gabions. When there are not enough boats to form a whole bridge over a large rapid river recourse is had to flying bridges, *e.g.* those over the Rhine.

Casks are easily procured and are a good material for rafts. They support infantry four deep, and field-guns of cavalry in file. The bearings should be 18 feet 9 inches (*vide* Fig. 115).

Piers of casks. * The method of forming a pier with casks is as follows :—

The party for each pier consists of 2 N. C. officers and 16 men, drawn up two deep with their right resting on the water. They are numbered from 1 to 8 by files. Two piers when launched are formed into a raft.

The stores required for each pier are :—

Seven casks, 2 gunwale pieces, 2 slings, 2 braces, 2 painters. 1 boat-hook. The men bring the stores down to the place selected—Nos. 1 and 8 bring the gunwales; Nos. 2, 3, and 4, the front rank man of No. 5 bring each a cask ; the rear rank man of No. 5 brings two slings ; Nos. 6 and 7 the braces, painters, and boat-hook.

The casks are laid out in one row side by side, bungs up and touching at the swell. If there should be any perceptible difference they should be sized from flank to flank. The slings with their loops to the front are laid down close to the ends of the casks. Each man takes a brace, except the gunwale man.

At the command "*Form Pier*" the men take post * facing inwards, front rank on the left, rear rank on the right of the casks. * Drill.

At "*Line Casks*" the N. C. officer superintending causes the men to arrange the casks with greater care, and numbers them, the nearest to himself being 1. If the casks are of unequal length they are dressed by the centre, so that the larger ones project equally on both sides. When properly arranged, all the men hold the casks steady except the gunwale men, Nos. 1 and 8, who give up the end casks to Nos. 2 and 7, and step back a little.

At "*Gunwales*," the gunwale men lift the gunwales over the heads of the other men, and place them along the line of casks on each side, not more than 6 inches from their ends, and with the centre of each gunwale corresponding to the centre of the middle cask, in respect to which they must take their direction from the man opposite to that cask.

At "*Slings*," No. 1 takes up the loop ends of the slings, and passes them over the ends of the gunwales. The cask numbers pass the slings under the casks,

No. 8 at the other end, setting one foot against the end cask and hauling taut.

No. 8 then brings up the ends of the sling inside the gunwales, takes 1 or 2 turns round, and makes fast with a clove hitch. Meanwhile the men steady the casks, and keep the slings in their proper places immediately under the gunwales.

At "*Braces*," the gunwale men steady the piers. Each of the others, with the eye of a brace in his hand, passes it round the gunwale. If there are no eyes to the gunwale, he makes fast with a timber hitch, and passes the brace outside downwards, and under the sling, exactly at the centre of the interval between 2 casks. The whole then mount upon the gunwales and haul taut.

At "*Taut*," each man dismounts, still holding taut, takes a round turn round the gunwale on the left of the standing part (which should be quite perpendicular), and holds on.

At "*Cross*," each man holding the brace in his left hand close up to the round turn passes over the end round the standing part with his right hand to the opposite man by his own left, and lays it down in the interval between the casks. He then takes his comrade's rope, passing it round the standing part of his own brace on the left side, while the opposite man does the same. They again change ends, but by their own right.

At "*Rock and Heave*," the men place their left feet against the gunwales, and rock the casks, hauling on the braces with both hands.

At "*Frap and make Fast*," they take another turn round the gunwales by the outside, and on the left of the standing part of their brace. They then frap the parts of the rope which embraced the gunwale, and make fast with a clove hitch round half the returns.

When properly done, two round turns lie evenly on the gunwales without riding, and the braces pass each other diagonally without crossing, each man's brace being on his own left of the brace opposite at the part where they cross.

Whilst the men are employed making fast the braces, 2 gunwale men stand by the end casks, while the other 2 get ready the painters.

At "*Painters,*" one is fixed at each end to the opposite gunwales, or to the slings immediately below the gunwales.

It may here be remarked that when men first commence this work, instead of launching the pier, which is the next thing to be done, they should be practised in taking it to pieces, and putting it together again, until perfect, when the piers are formed in the shortest possible time. * Repetition till perfect.

At "*Dismantle Pier,*" each man undoes, clears away, and coils down what he has done or brought up.

For launching, at the command "*Bring up the Ways,*" the gunwale men bring them up, attaching the 4 way-ropes if necessary which are fixed to the end eyes, and should be about $3\frac{1}{2}$ fathoms long, according to the side the ways may be on. * Launching.

At "*Prepare to raise the Pier right*" (or *left*), the gunwale men stand by the ways, the others raise the pier as directed. The gunwale men place the ways so that the bulge of the casks shall be in the centre of the ways.

At "*Lower*," the men lower the piers onto the ways.

At "*Man the Ways*," the whole of the numbers man the ways, and launch the pier. If formed on the ways it will be ready for launching as soon as completed. All man the way ropes; No. 1 holding on to the painter; No. 8 to the way rope.

At "*Form Raft*," each rank of the pier party falls in two deep, and Nos. 1, 3, 5, 7 front rank, 2, 4, 6, 8 rear rank. The commander and No. 7 bring a pier close to the shore, and make fast the painters to pickets fixed for the purpose. Nos. 2 and 8 bring each a transom and 4 lashings, and with Nos. 1 and 7 lash the transoms to the gunwales near their ends.

If the bungholes of the end casks should be covered by the transoms when the casks are upright, they must be turned a little inwards, so as just to clear the hole, and allow a hand pump being inserted to pump out any water there may be in the casks.

The command "*Baulks*" follows immediately, and all available men lay the ends of the baulks, 6 in number, on the gunwales, the distance between the two outer depending on the length of the chesses.

At "*Boom out*," the pier is pushed out by means of the transoms, and when at a convenient distance, the order is "*Hold on.*"

At "*Second Pier*," the commander and 7 men bring up another pier parallel to the first, the transoms and baulks being raised up, so that they may pass under them. Nos. 2 and 8 lash the transoms to the near pier, and the outer baulks to the gunwales.

At "*Chesses*," the commander and 2 men place them, the remainder hand them on board, together with rack lashings, anchors, boat-hooks, &c. If necessary, the transoms, baulks, chesses, and ribands for the bay are also handed on board; the chesses being piled up; the baulks laid outside on the gunwales, transoms outside the baulks.

After the chesses are laid down, the ribands are lashed down, and a turn of the lashing taken round the outer baulks.

To form bridge from a raft in smooth water, the rafts are alongside the beach in order, 1, 2, 3, &c., below the point where the bridge is to be formed. A boat party and a shore party are required to carry over the sheer line, to lay out anchors, and to complete communication with the shore. * Forming bridge.

At "*Form Bridge*," the shore party haul up No. 1 raft to its place, make fast painters, hand on board the shore transoms, baulks, and chesses. The commander of the raft lays hold of the shore line. Nos. 1 and 7 lash transoms.

At "*Haul out*," the commander makes fast the shore line. No. 2 raft is warped up alongside, and hauled into its place close to No. 1. All the crew of No. 1 hand baulks and transoms to No. 2. Nos. 2 and 8 of No. 1 raft lash transoms. Nos. 1 and 7 of No. 2 raft lash transoms.

The commander of No. 2 raft makes fast to the shore line, and all the numbers of No. 1 pass chesses and then rack down. Nos. 3 and 4 rafts follow in order.

The boat party place the off-shore baulks and transoms, and make fast painters, &c. As soon as each raft is ready, the crew sit down on the edges of the roadway, with their feet over the sides—even numbers on the starboard side, odd numbers on the port side. A second sheer line may be added if necessary. If the bank is very shelving, a temporary wharf may be required on each side.

Dismantling. * To take the bridge to pieces, at the commands— "*Attention,*" "*Prepare to Dismantle,*" the crew of each raft and the shore party withdraw the chesses off the bays.

At "*Dismantle,*" Nos. 2 and 8 and 1 and 7 unlash transoms and hand them in. The other numbers withdraw baulks. The commanders cast off sheer-line lashings; and the rafts, by warping or otherwise, regain their original positions down stream.

Pontoons. Metal cylindrical pontoons are used in our service; all other nations employ bâteaux. Pontoons should satisfy the following conditions:—

They should form a stable bridge at all times; and their equipment should be simple, easily transported and repaired.

The bearing power of a bridge is reckoned at so many foot-pounds per linear foot.

A bridge, calculated to carry infantry four deep, will carry cavalry in file and field artillery.

Pressure of infantry, cavalry, artillery. Infantry marching four deep produce a pressure on the bridge of about 200 lbs. per linear foot; if crowded, 560 lbs.; cavalry, 233 lbs.; crowded, 350 lbs.; an Armstrong, 12-pounder, 462 lbs.; a 40-pounder, 888 lbs.

Cylindrical pontoons. The bearing power of a cylindrical pontoon decreases after it is immersed half its depth, whereas in boats the bearing power increases until the whole is submerged.

These pontoons have baulks 14 feet long, 6 of

which support the flooring. The roadway is formed of chesses 11 feet 5 inches long. One waggon carries 2 pontoons and stores for 1 raft. Its weight is 3,425 lbs. Cylindrical pontoons are difficult to repair inside, are unmanageable in rough water, and it is not easy to remove water out of them.

The Belgians use an open boat 24 feet 6 inches long, square at one end, of sheet iron. 2 such boats, keyed stern to stern, form one bâteau, and can support 584 lbs. per running foot. Bridges are made of single or double bâteaux. The roadway is formed of single planks 9 feet 6 inches long. A waggon carries one boat and the superstructure for one bay. Total weight, 3,582 lbs. Belgian bâteaux.

The Austrians use an open boat of wood, in 3 parts, which can be used 2 or 3 together. The baulks are 23 feet long. The roadway is formed of single planks 9 feet 6 inches long. Weight of waggon load, 2,356 lbs. Austrian bâteaux.

The Belgians and Austrians approve of long bearings, and employ heavy baulks carried by 2 men. With them additions are made outside to form bridge. With us a pontoon bridge is formed by booming out from the shore.

The French have a boat of wood, weighing 150 cwt., with a roadway 11 feet wide, and bearing of baulks 19 feet. French bâteaux.

When anchors cannot be cast, a sheer line is necessary, and 3 boats are required to place it. One boat attends to the line in the centre, which is attached to a buoy; the other two carry the ends to the shore. Sheer line.

To protect a bridge from fire-ships, or other

Protection of bridge.	dangerous floating bodies, a boom of timbers or chains must be made strong enough across the river above the bridge, or a boat-guard established to prevent the approach of anything dangerous.
Advantage of continental system.	This shows one advantage of the continental boat system, for the boats are available for this purpose as well as for supporting the bridge. In passing military bridges, troops should not keep time.
Belgian trestles for getting onto a bridge.	It is sometimes difficult to get onto a bridge, owing to the steepness of the banks, or the uncertain level of tidal rivers. The Belgian trestles allow the roadway to be raised or lowered by screws, and are useful in such a case, or cask rafts may be employed.
Bridge * boats over the Adour, 1814.	Wellington constructed a remarkable bridge across the river Adour in 1814, previous to his attack on Bayonne. The difficulties to be overcome were— the river was large, tidal, and subject to heavy swells. The bridge, therefore, had to be flexible to allow for the rise and fall of the tide; 48 large coasting vessels were got over the bar with difficulty, and moored 40 feet from centre to centre. Five 13 inch cables were stretched from shore to shore, resting in notches cut in sleepers. One end of each cable was fastened to an 18-pounder, thrown over a 14-foot wall bounding the river on one side; the other ends were stretched taut by capstans and guy tackles, fixed to a frame of timber laid on the sand behind the other wall.

A boom of 2 rows of masts was constructed to protect this bridge from fire-ships. It remained uninjured till the close of the war. The success of this operation was partly owing to the uncertainty into which the French were thrown by the posting of a battery of six 18-pounders higher up the river.

From this instance, and many other passages of rivers in presence of an enemy, it will be seen that to deceive the opponent as to the point of passage is of the highest importance. In this case a few troops with a battery of rockets were pushed over to clear the opposite bank. *Enemy to be deceived.*

In 1809 Napoleon crossed the Danube by a number of bridges, some of Austrian pontoons, some of rafts, some of piles—one remarkable bridge was formed in a narrow channel, towed down stream into its place, and allowed to swing with the current. Its total length was 141 yards; but to allow of its moving in a winding channel, it was constructed in 4 parts, loosely lashed together, which the pontoneers secured, when they got it into position. The swinging took 3 or 4 minutes. *Passage of the Danube, 1809.*

Some waggons made of galvanised sheet iron, first used in America, have been adopted. They are easily repairable, and can be employed as waggons or pontoons. 4 make a raft. The weight of each is 5 cwt., and of the waggon complete 17 cwt. If 16 men are in one body it will sink 1 foot. They do not answer in a current. *American sheet-iron waggon.*

Rafts are easily constructed of any floating bodies, and cannot be sunk by artillery. When made of trees the ends should be tarred, if they are to remain long in the water. Rafts of 2 tiers of trees are secured together by boring through the upper tier, and partly through the lower, and driving in wedges. They are frequently used on the Rhine. *Rafts.*

P

Rafts can be made of hides. Untanned hides are used in India over a framework of bamboo.

Wooden trestles. * The simplest support for a bridge over a shallow river is trestles. The bottom must be firm, and the current not rapid. Julius Cæsar crossed the Rhine by their means. The early form of trestle consisted of 4 legs, supporting a ridge-piece; the legs spread out in two directions, and are secured together by diagonal braces of wood, as shown in Fig. 116. If carried with an army they may be keyed together; if made on the spot they may be fastened with wooden pins or tree-nails.

Crib-work. * In 1811, in the Peninsula, trestles were made to contain between their legs a considerable weight of well-packed stones, secured by strong planks in narrow-pointed cases, to increase as little as possible the breadth and section, and the resistance to the current.

Sloping piles supported the trestles, and horizontal chains lying at the bottom were made fast to strong piles driven above the trestles. The river crossed was 396 feet wide.

Bridge across the Prah, 1872 * A bridge of crib-work was made in 1872 to cross the river Prah in Ashanti, timber being plentiful and nothing else at hand. A tray at the bottom can be filled with stones, and sunk, when towed into position (*vide* Fig. 117).

Crossing of the Beresina, 1813. On the retreat from Moscow in 1813, Napoleon, with the wreck of his army, crossed the river Berasina on two trestle bridges at a point where the river was 100 yards wide and 7 feet deep, with a muddy bottom and moderate current, but much loose ice drifted down.

Fig. 117.
CRIB.

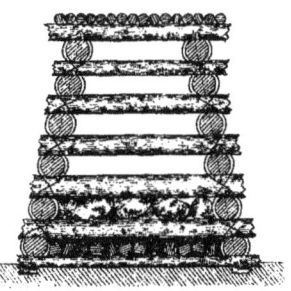

Fig. 118.
FLYING BRIDGE.

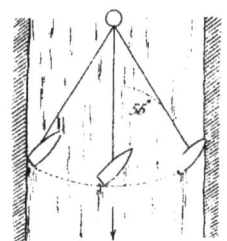

Fig. 119.
RAFT FOR FLYING BRIDGE

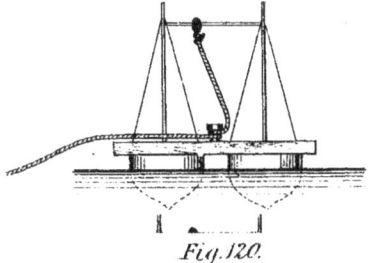

Fig. 120.
TRAIL BRIDGE.

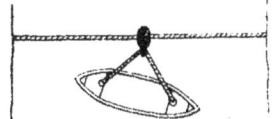

The trestles were 3 feet to 9 feet high; and the length of the ridge beam was 14 feet. 23 trestles were used, making 24 bays of about 13 feet each.

Trees about 16 feet or 17 feet long were used for beams in the round, there being no time to square them. Round timber 16 feet long, and 3 inches or 4 inches in diameter, formed the roadway of one bridge for carriages, which jolted greatly on so rough a surface, and shook the bridge violently. Old planks formed the roadway of the other bridge for infantry and cavalry. No small boats were at hand to assist, and it was difficult to keep the trestles steady; they sank unequally in the muddy bottom, and several were broken. The planks of the other bridge were unsound, could not be properly fixed, and were constantly being deranged: they split under the horses' hoofs or got into holes. When the Russians advanced four days afterwards, the bridges were destroyed by fire, and vast quantities of ammunition, artillery, and baggage were abandoned. " Thousands of men and many women and children were left to the mercy of the enemy and rigours of the climate."

The Belgian three-legged trestle is now chiefly employed. It consists of 2 cheeks, 14 feet long by 5 inches wide, by 3 inches deep, and a pry pole, connected at their heads by a pin, and kept asunder at their feet by bars 4 feet 6 inches long. A saddle or chapeau slides up and down the cheeks, and is supported at any required height by pins passing through the cheeks. Two such trestles are placed opposite, and a heavy cross-beam is laid on the saddles to receive the roadway. These trestles can be set on the bottom

Belgian three-legged trestle.

of a stream, or upon boats, and by their means a descent can be made from a high bank. They answer in a stream 12 feet deep. The cheeks of these trestles may be used for pile driving.

Lengthening bridge by trestles.
A bridge may be lengthened by the aid of a temporary trestle, consisting of 2 legs and a cross piece, in the absence of boats. The length of roadway between 2 sets of trestles is laid upon this and boomed out on rollers: The permanent trestles are then dropped and the temporary released.

Boats are used for crossing rivers as row-boats and flying bridges, besides forming bridges of boats proper.

Rowing-boats. *
Their employment as row-boats individually applies only to passages in face of the enemy, and for desultory operations. The difficulty is to get sufficient boats to pass the men over, and to return for supports; they are, also, liable to be sunk, thus causing confusion. The men ferried over must sit down. As few rowers as possible should be employed. The heads of the boats should be slightly up stream, so as to cross diagonally, aided by the current, and the best points for embarking and landing should be chosen.

Flying bridges. *
Flying bridges are common in many rapid rivers, the current being utilised to swing a boat carrying a load from side to side. The raft is sometimes attached to a boat anchored in the middle of the stream, and its side is then kept at an angle of 55° with the current (*vide* Fig. 118).

Communication was established across the Thames between Tilbury and Gravesend in 1778 and 1795, when invasion was threatened.

It is not easy to arrange such a bridge in a slow stream.

When the cable is fastened to an anchor, it is liable to be caught in the bottom. It is, therefore, necessary to support the cable on intermediate floats, and attach it to a mast. But for fear the raft is pulled over by the strain, it is better to form a raft of 2 boats together, with a traveller running on or in a crosspiece between the two masts; the cable is not permanently attached to the traveller, but placed on a capstan on deck to be taken in or let out as required. A greater load can be transported in this way than with a single boat (*vide* Fig. 119).

The two following points should be considered in all flying bridges :—

i. The cable should never be less than one and a half times the chord of the arc swung across, which is generally the width of the stream.

ii. It should be moored by two anchors, as when one only is used it is liable to be gradually worked out of its hold in the bottom.

A trail bridge is made by tightly stretching a * hawser across a stream, with a traveller on it, to which the boat is attached. By means of two ropes attached to the bow and stern and the traveller, the current can be utilised to assist in sending the boat across, the bows being kept in the required direction by hauling on or loosing the ropes (*vide* Fig. 120). The boat should be shaped like a whale boat, to go both ways. A flat-bottomed boat, with a traveller on a rope drawn by four negroes, conveyed foot passengers and carriages between George's Island and Hamilton in Bermuda before the construction of the causeway.

Trail bridge.

Crossing the Alva by Massena, 1811.

* Marshal Massena, in his retreat from Portugal in 1811, established a rough and ready communication across the river Alva. Ney had previously destroyed the bridge. No materials suitable to make good the broken arch could be obtained, therefore the passage was made at an adjoining part of the river where it was narrower and more rapid. For this reason, and because of the great depth, no supports could be fixed in the bottom. The right bank was bounded by a stone wall 5 feet above the surface of the current; the left was rocky, and nearly level with the water. The arrangement is indicated in Fig. 121.

Two ring-bolts were let into the rock at D and C. Six pine-trees 60 feet long were procured, and the raft G made ready near C, under which beams were placed on which to launch it. One of the pine-trees was then secured to the raft, the other end being fastened by a heel-rope to the ring-bolt D. A strong party held the rope E F, and the raft was let down into the water and eased off until it assumed the position at G, the end of the rope being made fast to the ring-bolt at C.

A firm footing being thus established in mid stream, the end of another tree was slipped along the first tree to G, and made fast to the raft. The other end was secured to a large cask. This was launched, and pushed out from I to K, whence, being caught by the stream, it was gently eased off by the rope L M to the wall at O. 30 men were then sent across, who hauled the cask and end of the tree on to the wall. Two other trees were then laid parallel to the former, and the whole planked with doors, chests, &c.

Fig. 121.
MASSENA'S BRIDGE.

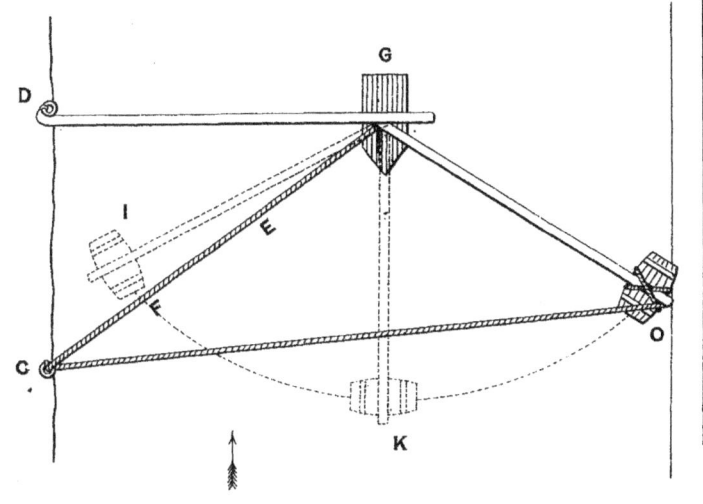

Fig. 122.
BRIDGE OVER THE COA.

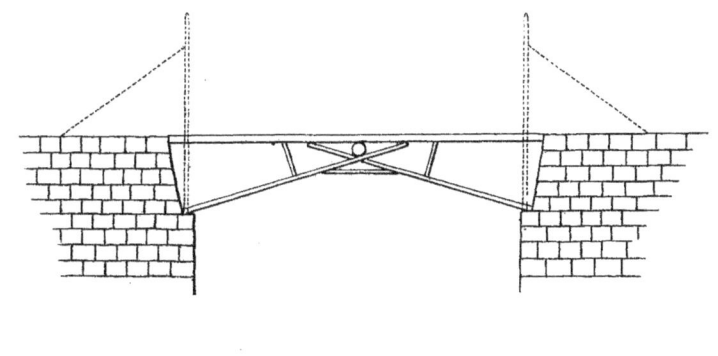

Another form of bridge was used at Almeida across the river Coa, the French having destroyed the stone bridge there in their retreat. Notches were made in the masonry on both sides to receive the lower extremities of two frames, which were eased down onto them from the edges of the gap in a vertical position. 2 tackles, applied to each frame, led to ring-bolts, set in the masonry, about 30 feet from the gap. 50 men being put on to each tackle, the frames were lowered down so that their upper extremities crossed each other. Gangboards were then shoved out, and men sent across to put in the key-bolts previously prepared. A ridge-piece being fixed in the fork, and beams laid across, the communication was complete (*vide* Fig. 122). * Bridge over the Coa, 1810.

Lever bridges are of rude construction, made by cutting down trees and sinking the butts in the banks sufficiently deep that the buried parts may exceed in weight those out of the ground (*vide* Fig. 123). * Lever bridges

Beams or trunks of trees should be laid over the butts, and secured by pickets, as in fixing abattis, then the earth well rammed above them. If the trees are not long enough to meet, so that their ends may be secured to those opposite, thinner trunks must be fastened to the ends out of the ground on one side, and bent down to meet the corresponding ends of the trees sunk in the opposite bank, to which ends they are lashed.

If a river is narrow, deep, and rapid, a tree may be felled on each side, and the branch ends gradually launched into the current, so as to join in mid stream. Men can then cross one by one. Often * Crossing by trees.

a single tree will be long enough to reach to the opposite bank, over which a party may cross to assist in preparing a bridge of one kind or other, according to circumstances.

Ladder bridge. * Fig. 124 shows how a ladder bridge may be formed by running a trench-cart or gun limber into the stream, and securing it there, with the shafts vertical, by ropes on both sides. Ladders are laid on the body of the cart or limber from both banks, and covered with planks. A similar bridge can be made by waggons placed lengthways in a shallow rapid stream. Such bridges support infantry.

Frame bridges. * Frame bridges are used generally to restore existing bridges which have been damaged or destroyed. They are of various kinds, made of frames such as shown by Fig. 125. The following is a description of a single lock bridge, of which the other kinds are modifications :—

Construction of single-lock bridge. * The section of the river or chasm must be accurately taken and marked on the ground with a cord and pegs, and the distance between the footings on one side must be 18 inches wider than on the other, otherwise the frames will not lock. The standards (*vide* Fig. 125) are laid out on the section, and marked with chalk at the places for lashing the ledgers and transoms. The whole are lashed together on both banks. The ledgers should be above the standards, about 2 feet from the butts, the transoms beneath. The frames are carefully squared, and the broader must be 18 inches wider throughout than the other. A ledger should be about 1 foot longer than a transom. The braces are then lashed. The transom of the narrow frame should be about

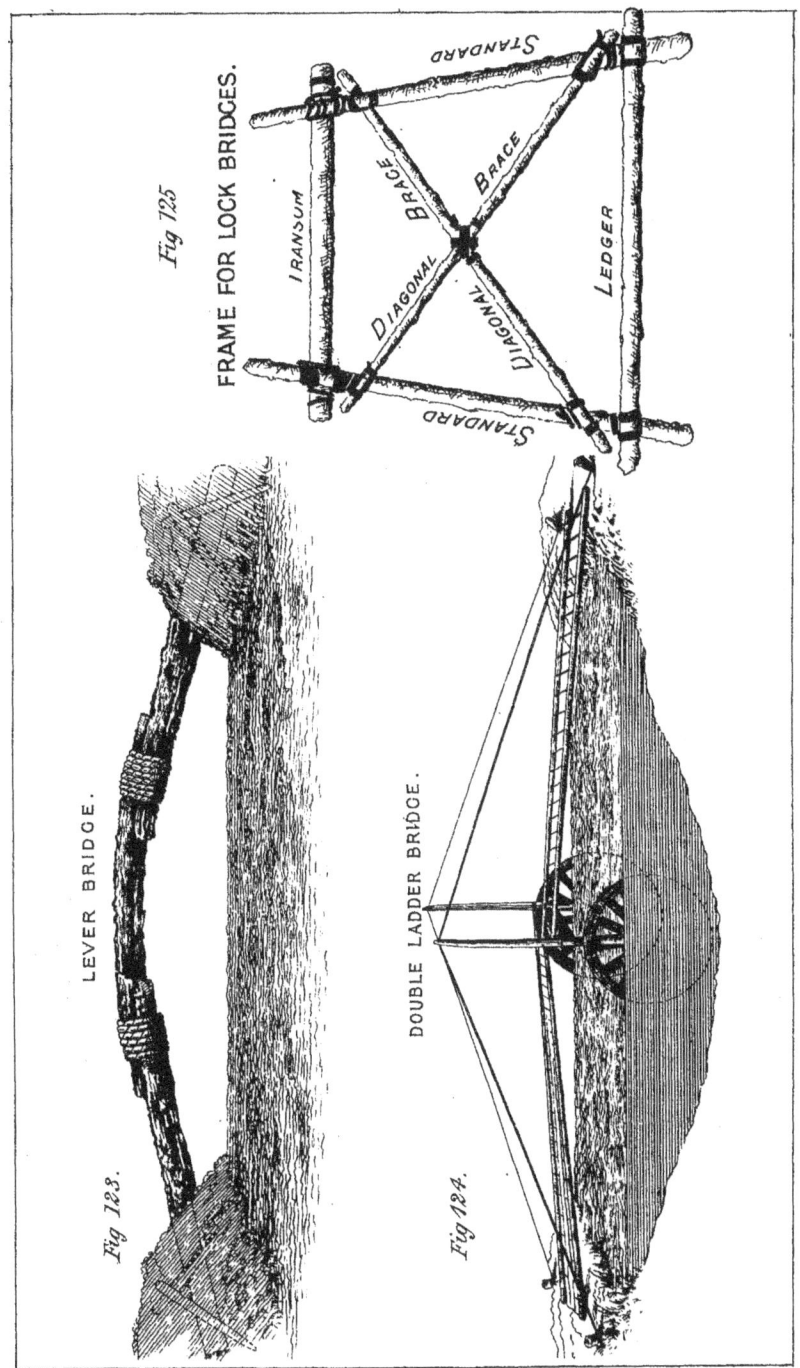

Fig. 126
SINGLE LOCK BRIDGE.

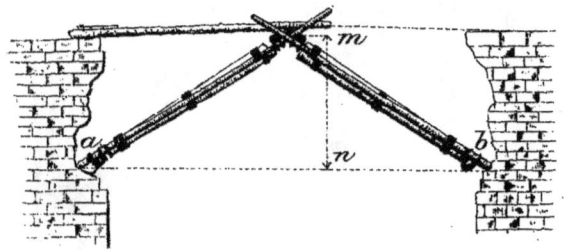

Fig. 127.
NETWORK OF ROPE BRIDGE AT ALCANTARA.

CONSTRUCTION OF BEAMS.

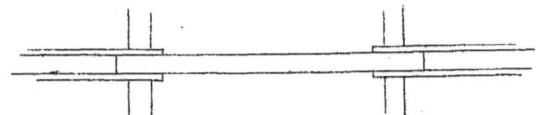

Fig. 128
PILE DRIVING.

18 inches wider than the width of roadway between the ribands.

Pickets are driven in for the guys and foot ropes. Foot ropes are fastened round the butts of the standards, below the ledgers, with a timber hitch. Fore and back guys are attached to the tops of each frame. The fore guys are crossed, those of the narrow frame passing between the horns of the broad frame.

The frames are launched and locked. Two temporary roadbearers are sent out, and 2 men working on the crutch of the bridge place the fork transom in position. The rest of the roadbearers are sent out resting on this transom. Chesses are laid and racked down, and a handrail rigged up.

The roadway is 9 feet wide, and will support infantry in fours.

The height mn should not be less than $\frac{4}{7}$ of ab, the span (vide Fig. 126).

The slope of the standards for these frames should be $\frac{20}{1}$, and for trestles $\frac{6}{1}$.

There are two sorts of suspension bridges. In one, the roadway passes over the suspending chains; in the other, it is hung from them. The former is the simplest, and one example, the iron gabion bridge, may here be noticed. Suspension bridges.

The iron bands of Jones's gabions are connected together by bolts and nuts. The bolts are $1\frac{1}{2}$ inches long and $\frac{5}{16}$ in diameter. The bands are looped round beams; one beam is held by tackle to the bank, and the bridge is stretched taut. The bands of 4 gabions, with the pickets woven between the bands, alternately over and under, will make a bridge of 40 feet to 50 feet span for infantry in file. * Iron gabion bridge.

A stronger bridge has been made of 6 chains, each chain formed of the bands of 4 gabions, 18 inches to 2 feet apart, over a span of 60 feet, able to support a field gun.

Rope bridge over Tagus at Alcantara. * A very celebrated bridge was made of ropes by Colonel Sturgeon of the staff corps in the Peninsula war to restore the communication across Trajan's bridge over the Tagus at Alcantara, to bring up stores from Badajos for the attack on the forts at Salamanca. The French had blown up the principal arch. The gap was 100 feet wide, and large enough timber could not be procured.

18 lengths of $6\frac{1}{2}$ inches cable, stretched across, were fastened by tackle. The cables were formed into a network by strong lashings, as shown in Fig. 127, and over this 8 cross beams were laid, having notches in their under sides to receive the rope. Other beams rested on the cross beams to support the planks, and were constructed as in the figure. The single beams were twice the width of each of the double ones. The junction was made by screw-bolts, allowing elasticity, and for the whole bridge to be rolled up and transported; for it was secretly made at a distance; and on 11th June, 1812, the siege train crossed by it.

Piles. * Piles are often useful for making piers, and to strengthen parts of a bridge. A convenient way of driving them by hand is shown in Fig. 128—two spars being lashed to the head of a pile, and a plank laid across them for the men using mauls to stand upon.

For temporary bridges, the *Handbook for Field Service* gives these rules as to strength of materials:—

5 deal battens 7 inches × 3 inches, supported at intervals of 15 feet, will carry infantry in fours crowded. _* Strength of materials.

5 larch spars 6½ inches in diameter, free from large knots and defects, will carry infantry crowded, or a 16-pounder.

Transoms should not be less than 9 inches in diameter, to resist the greatest strain to which a bridge is likely to be exposed, viz. that caused by infantry in fours.

The planks for a roadway to carry horses should not be less than 1½ inches thick.

To measure the breadth of a stream by the peak of a cap, stand on one bank and lower the peak until its edge cuts the other bank. Steady the head by placing the hand under the chin, turn gently round towards the most level part on your side, and notice where the eyes and edge of the peak again cut the ground. Measure this. It will be nearly the breadth of the stream. _* Measuring stream by peak of cap.

CHAPTER XII.

RECONNAISSANCE AND FIELD-SKETCHING.

Necessity for knowledge of a country.

"ON fait d'autant mieux la guerre que l'on connaît mieux le pays." So thought Napoleon. For the proper conduct of all operations in the field, a knowledge of the country is essential, as on it depend greatly the character of the movements and composition of the force. And it is truly said that—"Every manœuvre which is not founded upon the nature of the ground is absurd and ridiculous."

Meaning of a reconnaissance.

A reconnaissance may be described as the examination of part of a country, or of an occupied position, or a fortress, in order to obtain information as to its resources for supplies, and for movement of troops and carriages; also, as to the strength, position, and intentions of an enemy for purposes of attack or otherwise. If the object is specially to discover his strength, and make him develop his troops, we require an armed force. But if secret information of the enemy is sought, an officer of intelligence is usually selected, who approaches as near as possible to the position, and notes down all he can see and discover.

It is desirable that all N. C. officers should understand the rudiments of reconnoitring, for there are many occasions when information procured by a scout or a patrol may prove invaluable. There are some in every corps fully competent to furnish a reconnaissance of a road, river, &c., after a little instruction. But the duty generally devolves upon officers, who, either mounted or on foot, apply their knowledge of field-sketching to the best advantage usually with limited time at disposal. Hence, needless detail should be avoided, and great accuracy cannot be expected. Among the officers, also, there will be some quicker of eye, and more handy with the pencil than others, and they should be employed in preference. * Who should reconnoitre.

For field-sketching the reconnoitrer should have with him a prismatic compass, protractor, sketch-book about 9 inches square, with ruled paper, pencil, india-rubber, and a small box of colours. * Articles required.

Field-sketching or surveying is the art of ascertaining by measurement the shape and size of any part of the country, and representing it on a reduced scale in a conventional manner, so as to bring the whole under the eye at once. Field-sketching.

It is only the objects of tactical importance to which attention should be directed. These are hills, their steepness, relative height, the form of each feature; roads; all lines of communication; rivers; fords; marshes; woods; fences; and all other objects likely to affect the movements of the three arms.

A survey or plan is made with good instruments; the measurements being laid down at leisure, and the drawing executed with minute finish, when accuracy is more important than time. Survey distinguished from military sketch.

Whereas a military sketch is performed hurriedly, with simplest instruments, entirely in the field; and to become a good sketcher, practice in applying the principles of regular surveying is required, combined with rapidity and fair accuracy.

Hasty sketch. * It is possible to make a hasty sketch without any instruments at all, thus:—Having a sheet of paper or sketch-book, select the most convenient and level part of the ground as a base line, and step a few hundred yards along it, noting a tree, bush, or other object at its extremities. Draw a straight line on the paper corresponding with this, and mark off its length on the scale. Note from each end of the base the most prominent parts of the ground—*e.g.* hills, roads, churches, houses, &c. Place the paper on the ground, with the line drawn for the base, in the direction of the base, and without moving it mark off the angle each object makes with the base line and the approximate distance.

Meaning of * The scale is the proportion the sketch is to bear
a scale. to the ground represented. Scales are usually drawn to represent units, tens, hundreds, thousands of feet, yards, &c. The left division shows smaller quantities than are given in the other divisions, and these latter are numbered from the left division but one to the right.

A scale so divided is drawn with two straight lines, the under being thick, as shown in Fig. 129.

But when the divisions represent greater distances, as in scales of 1 inch or 2 inches to a mile, suitable for road sketches, &c., the scale is drawn with one line only.

The representative fraction of a scale shows the

number of real inches represented by one inch on the scale. It is also the proportion between a drawing made on the particular scale and the real object represented—*e.g.* in a scale of 1 inch to 10 feet the fraction is $\frac{1}{120}$; for 6 inches to 1 mile $\frac{6}{1760 \times 36} = \frac{1}{10560}$; for 3 inches to 1 mile $\frac{1}{21120}$. If a line a mile long be represented by a yard, the representative fraction is $\frac{1}{1760}$; if the same line be represented by a foot it will be $\frac{1}{5280}$.

Representative fraction.

The representative fraction should always be stated on a plan or sketch for this reason. An inch and a mile are measures applicable only to English operations, and do not convey any idea of the relative proportions between English and foreign plans, in which the same standards of measure are not used.

The scales usually adopted in field-sketches are 8, 6, 4, 2, 1 inch to the mile. The size depends on the nature of the ground, and the amount of detail required. Thus, for the sketch of a position, the scale may be from 6 inches to 12 inches; for road sketches, 1 inch to 4 inches.

The following are a few examples of simple scales :—

To make a scale of 6 inches to 1 mile to show furlongs—Divide a straight line 6 inches long into 8 equal parts, each of which represents a furlong. Divide the left division into 11 equal parts, each of which represents 20 yards; number the left division 220 yards on the left, and 0 on the right, and number the rest 1, 2, 3 . . . . &c., furlongs successively to the right.

Examples of simple scales.

To construct a scale of 6 inches to 1 mile to

read yards, we must calculate what length will represent 100 yards, thus:—As 1,760 yards are represented by 6 inches, 1,000 yards will be represented by $\frac{6 \times 1000}{1760} = 3\cdot 4$ inches. Divide a straight line A B of this length into 10 equal parts, each of which will represent 100 yards. To do this, from A draw a straight line A C, making any convenient angle with A B. Set off 10 equal parts on A C, join the last with B, and draw parallels intersecting A B— these will divide it into the 10 equal parts required. Produce B A, a distance equal to 100 yards; from A draw A D at any convenient angle, and divide the 100 yards into 10 equal parts in a similar manner.

If the scale to be made were larger, *e.g.* 24 inches to 1 mile, a straight line representing 1,000 yards would be of inconvenient length; and it would be better to take a less number, say 200. Then 200 will be represented by $\frac{200 \times 24}{1760} = 2\cdot 72$ inches. Divide a straight line of that length into 2 equal parts, to show hundreds. Take one of these parts on the left and subdivide it into 10 equal parts, to show tens of yards.

To make a scale of 1 inch to 8 feet. Here, 10 feet is represented by $1\frac{1}{4}$ inches. Divide a straight line of this length into 10 equal parts, each of which will represent 1 foot.

Suppose that a length known to be $2\frac{1}{2}$ miles is represented on a plan by 4·4 inches, construct the scale to read yards.

$$2\tfrac{1}{2} \times 1760 : 1000 :: 4\cdot4 : x$$
$$x = \frac{44 \times 1{,}000 \times 2}{10 \times 5 \times 1{,}760} = \frac{88{,}000}{88{,}000} = 1 \text{ inch.}$$

i.e. 1,000 yards are represented by 1 inch. Divide an inch into 10 equal parts for hundreds of yards. Add one of these parts on the left for further subdivision. The representative fraction is $\frac{1}{36000}$.

To construct a scale of 8 inches to 1 mile to read feet. Here such small divisions of 10 feet cannot conveniently be shown. So, since 8 inches represent $1{,}760 \times 3 = 5{,}280$ feet, 2,000 feet are represented by $\frac{8 \times 2{,}000}{5{,}280} = 3\cdot03$ inches. Divide a straight line of this length into 2 equal parts, each to show 1,000 feet, and add another part on the left for further subdivision. The representative fraction is $\frac{1}{7920}$.

To draw two plain scales, one to measure yards, the other paces of 30 inches, when the representative fraction of a plan is $\frac{1}{6000}$.

(1) $\quad \overset{12)}{\underset{500}{6{,}000}} : 600 :: 1 : x \quad x = 1\tfrac{1}{5}$ inches.

Divide a straight line of this length into 2 equal parts for hundreds of yards, and subdivide the left part to show tens of yards.

(2) 1 inch represents 500 feet = 200 paces. Divide an inch into 2 equal parts for hundreds of paces, and subdivide the left part to show tens.

If on a plan $3\tfrac{1}{4}$ miles are represented by $4\tfrac{1}{2}$ inches, to draw a plain scale of miles and furlongs long enough to measure 5 miles. Divide $4\tfrac{1}{2}$ inches into 13 equal parts, each of which will represent

Q

¼ mile. Adding 7 such parts gives a length of 5 miles. Bisecting each of the 4 left parts will show furlongs.

To draw a plain scale of $\frac{1}{9}$ to measure feet and inches up to 5 feet—$9' : 5' :: 1' : \frac{5'}{9} = \frac{60''}{9} = 6\frac{2}{3}$ inches. Divide a straight line of this length into 5 equal parts to show feet, and the left part into 12 to show inches.

To draw a scale of $\frac{1}{250}$ to show single paces, the maximum length being 60 paces, assuming the pace to be 30 inches. Here, 250 paces : 60 paces :: 30 inches : $\frac{180}{25} = 7\frac{1}{5}$ inches. Divide a straight line of this length into 6 equal parts to show tens of paces, and the left part into ten to show single paces.

If a plan has no scale attached, but some of the distances are noted on it, one marked 3 miles measures 6 inches. In order to make the scale, it is evident 2 inches represent 1 mile. Divide 2 inches into $\frac{1}{2}$, $\frac{1}{4}$, $\frac{1}{8}$ to show fractions of a mile, and add another 2 inches on the right for a complete mile.

To draw a scale showing miles and furlongs for a map on which the distance between 2 places, 35 miles apart, is represented by 8·75 inches. As 8·75 inches represent 35 miles, 9 inches represent 36, and 10 inches represent 40 miles; therefore $2\frac{1}{2}$ inches 10 miles.

Divide $2\frac{1}{2}$ inches into 10 equal parts for miles, and the left division into 8 equal parts for furlongs.

Comparative, diagonal, and other scales should be studied in text books.

Distances in military operations must not only be measured by length, but also by the time taken in traversing them. This varies greatly in mountainous districts, so that a scale of time, founded on experience, will be a most useful addition. Such a scale should show the hours over mountain passes as compared with the hours marching on level roads. <small>* Measurement by time as well as length.</small>

A map is easily copied on a larger or a smaller scale by drawing a series of squares on it, and another series on the paper for the copy, each square in the latter to be proportionately larger or smaller than the former. The detail of the map is then filled in. Thus, suppose a drawing with a representative fraction $\frac{1}{10560}$ is to be copied on a scale of 3 inches to 1 mile. The scale corresponding to $\frac{1}{10560}$ is 6 inches to 1 mile. Then, if we draw squares of 1 inch side on the original, the squares on the copy must be $\frac{1}{2}$ inch a side. <small>* Enlarging or reducing a map.</small>

Again, if a drawing on a scale of 3 inches to a mile, having $\frac{1}{2}$ inch squares marked on it, has to be copied so that 880 yards will occupy 3 inches on the copy. As 3 inches represent $\frac{1}{2}$ mile on the copy, 6 inches will represent 1 mile, and for copying the original, the squares must have sides of 1 inch.

The use of the prismatic compass can be understood in a few minutes. Every circle is divided into 360°. So the compass card under which the needle swings is graduated into 360° and half degrees; and <small>* The compass</small>

an angle can be taken to $\tfrac{1}{4}°$ by estimation if the instrument is perfectly steady.

To take a bearing. * To take a bearing, remove the cover, raise the sight vane, and uncover the prismatic glass. Keeping the compass level, so that the rim of the card does not touch the glass, raise it steadily to one eye, and bring the hair of the vane in line with the object. The number reflected up is the bearing, *i.e.* the angle a straight line drawn from the observer's eye to the object makes with the magnetic meridian. A reference to Fig. 130 will readily show the bearings of the four cardinal points, and the quadrant within which any intermediate angle falls, *e.g.* 45° is north-east. On windy days it is advisable to kneel, and the oscillations of the card may be checked by touching the spring under the vane.

Errors. * The compass is liable to error if near iron ore, or an iron gate, and is affected by magnetic disturbances. No two instruments read exactly alike, and accuracy can only be obtained to about half a degree.

Use of the compass in field-sketching. * In field-sketching, we determine the positions of objects by compass in four ways, viz. :—

i. By triangulation, which means fixing a certain number of conspicuous stations within the required area, by means of a carefully-measured base line, and the observations of the various stations from the two extremities of the base, or from stations previously fixed. Where the bearings intersect is the position of the object.

ii. By interpolation, standing at the object, taking the bearings of two or more fixed points from it, and laying these bearings off backwards from the

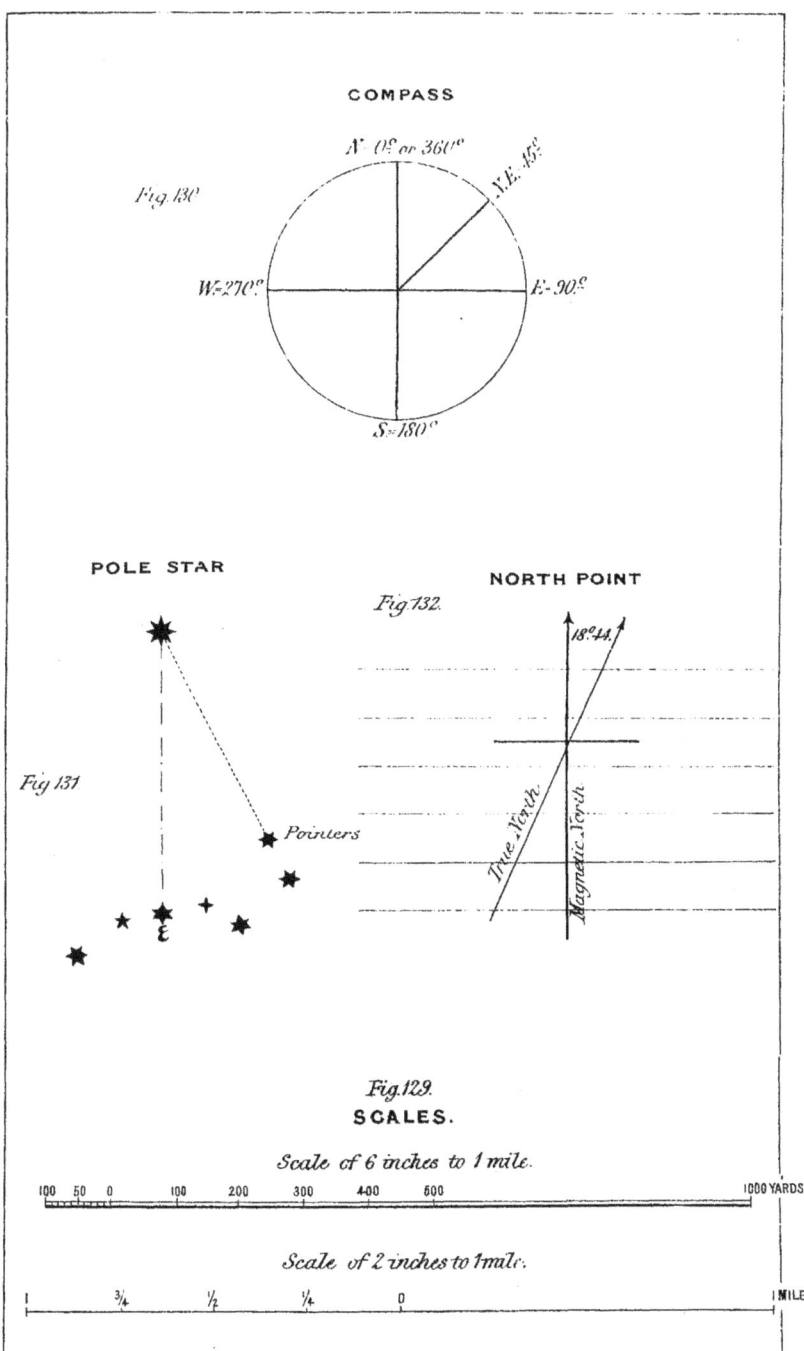

respective fixed points. The intersection of these bearings fixes the object.

iii. By taking the bearing of the object, and pacing to it along that bearing, laying the distance down to scale.

iv. By pacing from one fixed point towards another fixed point, and then pacing the length of the off-set.

The chief consideration in triangulation is the measurement of the original base, which should be as long as possible in proportion to the ground, and central—from 600 to 1,200 yards for a square mile; from $\tfrac{3}{4}$ mile to 2 miles for 10 square miles. It must also be level, or nearly so, to avoid reducing the measurement to the true horizontal length; and the ends chosen should have an unobscured view. The stations should be selected so as to afford good intersections, not too acute or too obtuse. An angle of 60° is best, so as to form equilateral triangles. The stations can then be determined with fair accuracy; and the base is not shortened for subsequent triangles. *Base.*

* Stations.

Interpolation comes into play when a certain object, e.g. a tower, is thought to be wrongly placed. Then, in order to correct its position, select any two points, the distance between which is known to be correct, and which are placed conveniently with regard to the tower, making the intersection of the bearings from 60° to 90°. This intersection fixes the correct position of the tower, which can be checked by the bearing from a third point. ** Interpolation.*

The "magnetic meridian" is the north and south line as shown by compass, deviating from the true ** Magnetic meridian.*

north and south line by a variation depending on the place, the year, and even slightly the time of day or night.

Variation from true north. * The variation is found by ascertaining the angle between the magnetic north and the true north, as laid down in the neighbourhood of an observatory, or from the bearing of the Pole star, when it is vertically over the star ϵ of the Great Bear (*vide* Fig. 131).

The variation changes constantly. In England in 1838 it was 24° 6′ W., since which date it has been going north again at the rate of about 7′ a year. From 1838 to 1884, the difference has been 46 × 7 = 322′ = 5° 22′; so that we may take the present variation to be 18° 44′.

Scale and north point. * No sketch is complete without a scale and a north point; the latter should be drawn as in Fig. 132, the magnetic north being at right angles to the horizontal lines on the paper, which run east and west.

Forward angle. By the "forward angle" is meant with the theodolite the horizontal angle between the zero line selected, and the direction we are about to proceed * in. With the compass it is the bearing of the line of direction in which we are pacing from the last station. The rule is to observe it last.

Back angle. * The "back angle" is the horizontal angle between the bearing of the forward station and the station just quitted.

Closing angle. The "closing angle" is the bearing taken of another station at the conclusion of a traverse begun from a previous station, to test the work.

Check angle. A "check angle" is the bearing taken from one or more subsequent stations in a traverse of some conspicuous object, not more than about ½ mile distant.

If the work is correct the bearings should converge at the point marking the position of the object fixed from previous stations.

An "offset" is a measurement taken perpendicular to the "forward angle." * Offset.

The traverse of a road, river, &c., is proceeding along it, taking bearings of different objects on either side at the starting point, and, last of all, the "forward angle"; then stepping the distance as far as the road is straight, where a halt must be made to take another round of angles, fixing distant points. All intermediate objects may be fixed by the eye. The distances of any buildings, bridges, hills, &c., met with on the road are noted in a field-book, which is often more convenient than the sketching-block itself, especially in bad weather. * Traverse.

This book should open lengthways, and have two parallel lines about ½ inch apart ruled through the centre of every page. The work commences at the bottom, and proceeds upwards. The parallel lines represent the right and left of the road; and all detail on or near the road should be entered neatly, and in conformity with the conventional signs. Other information is written. A clear system should be adopted, so that the work may be intelligible, if required to be transferred to paper later on. The distances along the road are entered within the lines. They are not noted from place to place, but in one count so long as the direction continues the same. As soon as the direction changes, as at a bend of the road, the total distance is closed with a circle for a new station, and a new count commences. Care must be taken not to confuse distances with angles, * Field book.

which should be written with the symbol for angles along a line right or left, with the name of the object or place to which a side road may lead ; and the "forward" angle should be distinguished thus—∠f 89°.

Plotting. * The process of laying down on the plan the work entered in the field-book is called "Plotting." For this purpose an ivory protractor is used, having parallel lines cut across it, and the whole of the angles of a semicircle marked on one edge, running into a point in the centre of the opposite edge, which is the centre of the circle. The manner of proceeding is very simple. Having fixed a point on the paper as a starting-point, in order to lay down a bearing,

Protractor. * first place the protractor to the right of the point if the angle is not greater than 180°, *i.e.* when it is in the eastern half of the circle, and to the left of the point when it is greater than 180°, *i.e.* when it is in the western half of the circle. Second, place the centre of the edge marked by a star exactly at the point. Third, make the parallel lines on the protractor coincide with the east and west lines of the paper. Then, holding the protractor steady, mark the angle on the outer edge with a point on the paper, and draw the bearing to it from the starting-point. The angles between 180° and 360° are marked on the protractor inside those from 0° to 180°, and every angle of each pair is the supplement of the other, *i.e.* the difference between them is 180°.

Laying off * This is useful, enabling an ungle to be laid down an angle backwards. backwards, as explained under the head of "Interpolation." Thus, if standing at a certain point we find an object already fixed on the paper bearing 55°, to lay it off backwards we take the angle 235°, and *vice versâ*. It has also another advantage, for when a point is so near the edge of the paper that the actual bearing cannot be laid down, marking the angle backwards and producing it from the first point gives the bearing required.

In pacing, precaution must be taken not to lose count. A good plan is to close a finger in the hand for every 100 up to 500, then to open the hand, mark down 500, and begin afresh. Although the regulation pace is 30 inches, yet most men step longer, especially when walking fast. Measurements are more likely to be accurate if we step yards and not paces; we thus avoid the trouble of reduction to yards to suit most scales, and we get over the ground quicker. Going up or down hill the distance will be longer than it is in reality; so we must step out. Other means for measuring may be used, such as a horse; but his paces at a walk and a steady trot must be carefully tested. And the circumference of a carriage wheel may be measured, and a rag tied to one of the spokes, so that the revolutions can be counted. It is always desirable to check the pacing by noting milestones or points, the distances between which are already known; and telegraph posts will be of assistance, when fixed at uniform intervals.

In the representation of slopes, a slope is considered the hypotheneuse of a right-angled triangle, and is represented by a fraction, the numerator of which contains the number of units in the perpendicular, and the denominator the number of the same units in the base, *e.g.* a slope of $\frac{1}{3}$ is one where the perpendicular is one-third the length of the base,

* Pacing.

* Other measurements.

Slopes.

while in a slope of ¾ the perpendicular is three times the length of the base.

Clinometer.
A clinometer is an instrument for measuring slopes and vertical angles. It can be made with a piece of cardboard about 6 inches long, on which a semicircle is described. From the centre a line is drawn perpendicular to the diameter, and is numbered 0 at the edge. Radii are drawn to cut the semicircle, reading units and tens of degrees. A plumb bob, suspended from the centre, shows the angle of elevation or depression at which the diameter is held.

Representation of hills.
Hills are represented on the system of contouring. Contours are imaginary lines drawn round a hill at equal vertical intervals. An idea of them is best obtained by considering the hill surrounded by water, the level of which rises first 10 feet, and the mark up to which it reaches on the different slopes will form the first contour. Suppose the water to rise another 10 feet, and the mark again traced, this will be a second contour, and so on.

Contours.
In order to determine the number of contours on a slope, we must measure the inclination, and pace its length, stepping out so as to approximate the horizontal measurement; then mark on a fixed straight line as many contours as there is space for horizontal equivalents, the scale of which is indicated on the protractor corresponding to certain slopes and scales.

Contours are usually shown at the vertical interval of 25 feet on a scale of 6 inches to 1 mile; of 12½ feet on a scale of 12 inches; and of 50 feet on a scale of 3 inches to 1 mile.

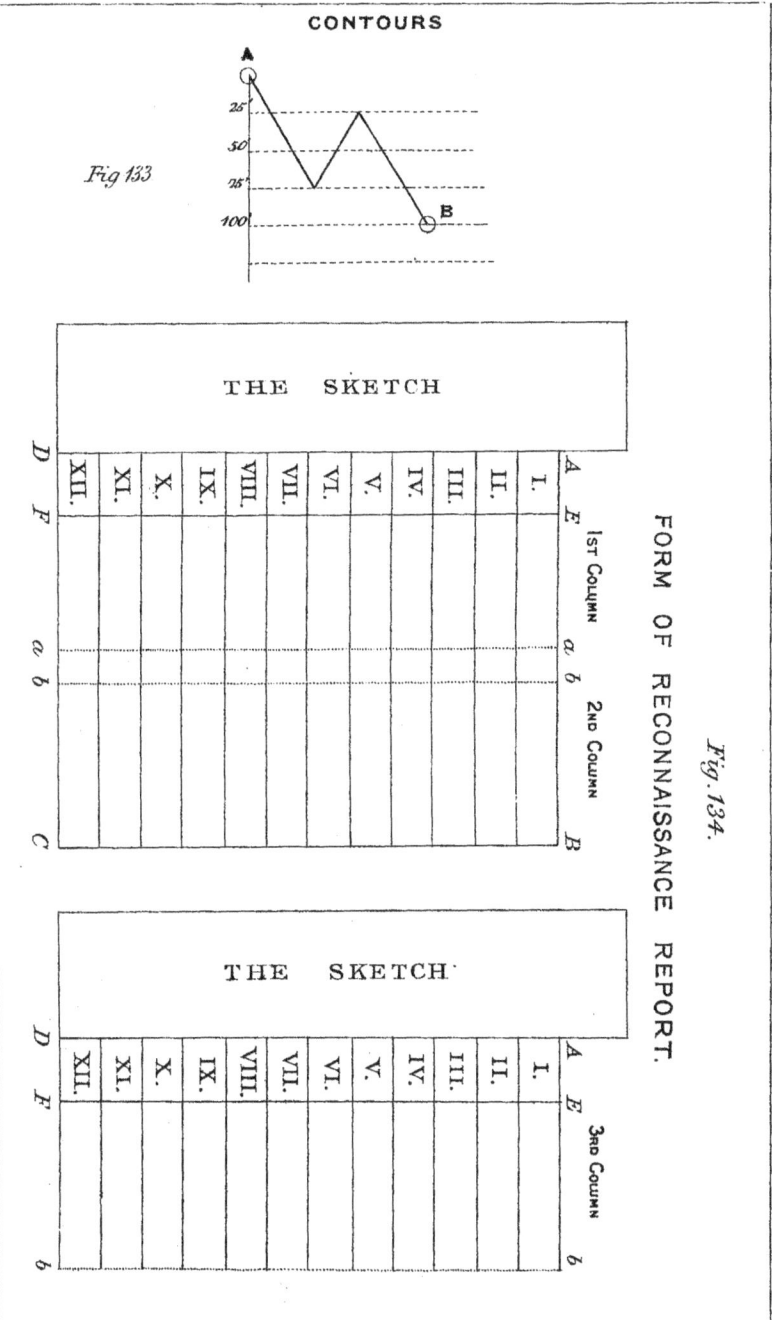

The horizontal equivalent of a contour is the base of a right-angled triangle, of which the perpendicular is the vertical interval between two contours, and the hypotheneuse is the measured slope.

For 1° for a 25 foot contour it is about 480 yards.
" " $12\frac{1}{2}$ " " 240 "
" " 50 " " 960 "

and for 2°, 3°, 4° . . . &c., we may assume it to be $\frac{1}{2}, \frac{1}{3}, \frac{1}{4}$ . . . &c.; these distances, *e.g.* if the contours are 25 feet apart, the slope 8°, and the measurement 600 yards, the horizontal equivalent will be $\frac{480}{8} = 60$ yards, and 10 such contours can be marked on the line up or down.

Horizontal equivalents may also be described as the horizontal distances between contours running at equal vertical intervals, being shorter as the slope of the hill is steeper. Thus, if a and b are two points immediately above one another on two contours, ac the horizontal equivalent $= bc$ (the vertical interval) × co-tangent $\angle bac$.

Suppose it is required to find the difference of level between A, the highest, and B, the lowest, point on a plan contoured with 25 feet vertical intervals, and there are 8 of these contours between these points, of which 6 show descending ground, and 2 show a rise, then a reference to Fig. 133 shows that the difference of level is 100 feet.

Time, however, will rarely admit of contours being run in detail; but the principle of them must be remembered in all field-sketching. The shading of hills may be in the vertical or horizontal styles, or in brushwork, which last is rapid and simple; but in every case the steeper the slope the more intense

Horizontal equivalent.

* Principle of hill shading.

must be the shading, and the lines thicker and closer together, while in a gentler slope they will be thinner and farther apart. The brows or salients will usually be found steeper than the valleys and re-entering parts. To represent slopes greater than about 35°, we must resort to vertical, broken strokes, such as are employed for rocks and precipices.

There is one disadvantage in a contoured plan, even when shaded, and that is, it requires often a practised eye to read the features correctly, for there is much similarity between the summits and the valleys; and this is specially so, when the surface has been gradually denuded, *e.g.* the chalk downs near Dover. Therefore, to make the drawing clearer, arrowheads should indicate whether the slopes run up or down; and the heights in feet of certain points should be noted.

Relative heights. * When a sketch is completed, the simplest way is to estimate two or three of the highest points, then of intermediate points, and of the lowest points, which last will be usually water; mark them 1, 2, 3, respectively, and state in the references their relative heights.

Aneroid. A pocket aneroid is very useful in determining heights; there is generally a small table in the case, and when the readings are taken at the base and summit of a hill, the height is found by deducting the former from the latter.

Or with a barometer, the difference between the readings at the base and summit multiplied by $\frac{10000}{11}$ gives the height in feet; $\frac{1}{440}$ of this result must be added or subtracted for every degree of temperature above or below 55° Fahr.

The greatest practicable slope for horses is $\frac{2}{5}$; for mules $\frac{1}{2}$; for men $\frac{4}{5}$. Baggage waggons ascend a slope of $\frac{1}{15}$ with difficulty. A slope of $\frac{1}{7}$ is impassable for them. Infantry can ascend $\frac{1}{3}$ or 20°; but for single soldiers steeper ground is practicable. Cavalry are at a disadvantage on a greater slope than 5°; and are useless on a steeper slope than 10°. Artillery finds 10° difficult, and is quite inefficient for action on a greater slope than 12°. *Practicable slopes.

The following is the authorised method to be adopted for all reconnaissance sketches and written reports:— *Authorised form of sketch and report.

The sketch is to contain all necessary information, which can be put in without confusion and without unduly increasing its size.

The written report is to be supplementary to, and explanatory of, the sketch, to convey necessary information with the utmost conciseness, and to be written on the form shown in Fig. 134. The form is folded so that the marginal headings may be seen in their proper places when the report is turned over for the perusal of the third column.

Before writing the report, every place referred to in the sketch should be marked with a red numeral, surrounded by a circle, consecutively upwards from the bottom. But places named on the sketch need not be numbered.

Remarks on a sketch should be made in the margin, clear of the drawing, connected with the points referred to by light lines.

The lower edges of both sketch and report are to be in line, and the sketch attached to the left margin.

Report. * In Figs. A, B, C, D, is an ordinary double sheet of foolscap open; aa is the original fold down the centre, and it is refolded along bb, a line parallel to aa, but a little to the right of it: thus A, E, D, F, form a margin visible when the report is turned for the perusal of the 3rd column. This margin contains the headings of the report in Roman numerals, I., II., III., &c.

Horizontal lines are drawn opposite each marginal heading to contain information concerning that heading, and these are continued over the page so as to form a 3rd column if necessary, and drawn so as to fall in their proper places exactly over the lines in the 1st and 2nd columns, when the page is turned down. First fill in the spaces in the 1st column, and then, when necessary, use the others.

The report is headed and signed at the back of the page containing the 1st column.

Reconnaissance of a road. * The various points to be noticed in reconnoitring a road are:—

i. The road itself—general direction; nature as hilly or otherwise; gradients over 5° should be clearly shown. Heavy waggons can scarcely pass slopes greater than 8°, unless with extra horses. Such a place, or any bad part of a road, may seriously delay a march, and should, if possible, be avoided.

Its construction and present condition; whether metalled or not; where materials for repair are to be obtained.

Whether sandy and heavy in dry weather, or of stiff clay and deep in wet weather; if drained or not. Quarries near at hand, gravel pits, and stone walls, furnish means for repair; fencing, timber, logs

CROSSINGS OF ROADS

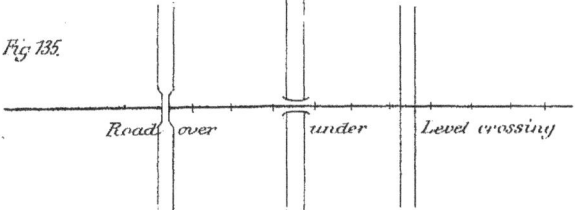

Fig. 135.

RICK OF HAY

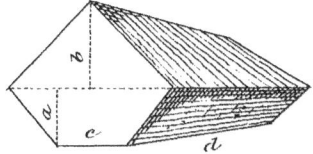

Fig. 136

RICK OF STRAW

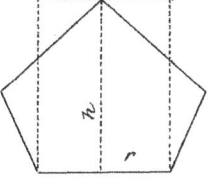

hurdles, small trees, brushwood, heath, planks, &c., should be mentioned if likely to be needed.

Width—which should be 20 feet for convenient passage of troops. Any narrow part should be specially noted. Guns in column of route require at least 3 yards, but more if there are sharp turns in the road.

ii. Bridges and fords.—A rough hand sketch should be made of any bridge showing dimensions, number, and size of piers; width and number of spans; thickness of arches at crown and haunches. We should mention, also, the material the bridge is built of, the length and width of roadway, the quickest way to destroy and repair it, if materials for repair are close at hand, and the nearest point of crossing if it is destroyed. * Bridges and fords.

The nature of the river crossed; what fords exist, to be used either in conjunction with the bridge, or instead of it if destroyed.

The position of the bridge, whether in a straight reach of the river, or at a bend, salient, or re-entering to the enemy. The approaches; adjoining country; means of defending it. (For fords, see *ante*, p. 155.)

When a bridge passes over a road, notice the height above the road, and if loaded waggons can pass under it.

It is particular to indicate clearly on a sketch when a road crosses another, or a railway, which of the two is uppermost (*vide* Fig. 135).

iii. Rate of marching, specifying whatever is likely to delay it. Length of defiles and hollow ways; * Rate of marching.

where danger chiefly exists. Note everything that will enable an officer to take defensive measures, and to guard against surprise.

Towns and villages. * iv. Towns and villages.—State the situation as regards the road; if adapted for fighting purposes, or for billeting troops; if inclosed by walls, fences, hedges, banks; built of stone, brick, or wood; tiled or slated roofs, or thatched; the position and number of entrances; chief substantial buildings; the character of the country surrounding; the streets, whether narrow and winding.

Count or estimate the number of houses, and allowing half for the inhabitants; the remainder will serve for accommodation.

We can allow 1 man per yard in a room 15 feet wide, and 2 men per yard in wider rooms. A rough estimate of the population can be obtained by multiplying the number of houses by 5. Ascertain their trade and occupation. If agricultural people, they will be better with the use of pick and shovel than if manufacturing. Inquire the market days, and where are the chief stores, bakeries, factories, butchers' and provisions shops, wheelwrights, forges, mills, telegraph office, &c.

Water supply. * v. Water supply, whether from rivers, ponds, wells, pumps. To water horses from the edge of a stream, the level of the water should not be more than about 4 inches lower than the bank. One well of ordinary depth will supply a battalion; but if the depth is very great it will take some hours to draw water. The number of horses which can be watered at a time at a certain part of a stream is the same as the width of that part in yards. A horse requires about 3 minutes to drink. It is advisable to dam up a small stream.   * Watering places.

vi. The nature of the country, hilly or level; open or wooded; pasture or cultivated; what trees form the woods, if with tall stems, comparatively open, or thick with brushwood and low branches. Notice if the fences are large or small, what walls and ditches exist affording cover or obstacles—any high places near the road suitable for signal stations—any dry spaces of grass or heath suitable as halting-places, clear of a village, with water convenient; or shady places in warm weather.   * Country.

vii. Rivers.—Note the following points, which apply more fully when the reconnaissance is of the river itself:—The force of the current and its direction, shown on the sketch by an arrow-head; all bridges, ferries, and fords, with roads leading up to them, crossing or running parallel to the river:—its breadth and depth; whether liable to floods, and the extent of country flooded; the character of the banks, their height, and which commands the other. The number of arches, span, and height of bridges; depth of water under the bridge; any commanding ground within range of field guns; boats and barges available; islands; towns, villages, or detached houses on the banks; whether the river is navigable with facility or not; whether tidal; how high up the tide extends, and its average rise and fall (*vide* Crossing Rivers, p. 198).   * Rivers.

viii. Camping ground, whether desirable, on gravel or on clay, or if the country is low and marshy; affording sufficient supply of good water—5 gallons   * Camping grounds.

R

per man and 10 per horse a day is ample; if fuel is procurable (*vide* Encampments, p. 8).

Positions. * ix. Positions, such as would be advantageous to an advanced or a rear guard, should be particularly noticed, *e.g.* cross roads, a ridge of hills commanding the road, villages, woods, and broken ground, keeping constantly in view the object of covering our own troops, and of causing the enemy to traverse an open space under fire.

Railways. * x. Railways, where they lead to, the gauge, single or double track; whether on an embankment, in a cutting, or on level ground; position of stations, size, length of platforms, convenience for transport of troops; rolling stock and stores: tunnels, bridges, viaducts; whether it is possible to march troops along the line. Further particulars would be obtained by a special examination of the railway.

Supplies and accommodation. * When a reconnoitrer is directed to obtain information of supplies and accommodation in a district or a separate place, it is desirable to tabulate it in some such form as the annexed :—

Name or reference number of place.	Distance in miles and furlongs.		Population.		Accommodation on a march.		Supplies.	Transport.	Water.
	Intermediate	Total.	Houses.	Persons.	Men.	Horses.			

Crops. * Regarding crops, an acre yields about 25 bushels of corn, 45 of oats, 35 of barley, 25 of peas, 30 tons of turnips, 15 tons of carrots, and 8 tons of potatoes.

Bricks. * A brick is 9 inches long, 4½ inches broad, 3 inches thick, and weighs 9 lbs. An unskilled man can lay

300 to 500 in a day; a very good workman 1,000. A man can wheel 30 bricks in a barrow.

Men extended 5 paces apart can clear 30 paces forward of brushwood 5 or 6 years old in 8 hours. * Clearances.

Sometimes tall grass, corn, and reeds, impeding the view, should be cut or trampled down or burnt. Men 2 paces apart can cut down hedges in from 6 to 18 minutes, pulling down the top branches with a pole and rope, and if necessary cutting the lower branches with an axe.

There is no better little work on reconnoitring than that entitled *What to Observe and How to Report It*, by Colonel Hale, Royal Engineers. Every sergeant should possess a copy. That officer classifies reports in three kinds—1st, verbal; 2nd, written; 3rd, drawn. Sometimes they are of two or all combined. * *What to Observe and How to Report It.*

1st. A verbal report is one made by word of mouth. The soldier making it is advised not to be, and not to speak in, a hurry; to settle exactly what he is going to say; and if he has made a mistake to say so at once. * Verbal report.

2nd. And because it is difficult to give a verbal message accurately, the soldier is recommended, when possible, to write it down on a piece of paper, no matter how small or untidy. He should write clearly, read aloud what has been written to detect mistakes, sign his name, rank, and regiment, and put down the minute, hour, and date; for what is true in one month may be incorrect in another, or even what is true in one hour may not be so in another—*e.g.* a river may be fordable in June and a * Written report.

torrent in January; at 6 A.M. it was crossed on foot, but at noon the tide has risen, and it cannot be crossed on horseback.

Report drawn and written.
* 3rd. The third kind of report is a drawing, without which it is often difficult to explain the position of roads, houses, woods, troops, &c. Certain marks, called conventional signs, are made use of to denote different objects. They are, as it were, the alphabet of the reconnoitrer. All employed in reconnoitring

Conventional signs.
* should use the same signs. Plate X. shows the conventional signs authorised to be employed.

Avoid repeating information. What is clearly shown in a sketch need not again be described in a report.

Reconnaissance of a position.
If an officer is directed to reconnoitre a position his attention should be given to the advantages and disadvantages it offers, the neighbouring country, the communications, rivers and bridges, crossing them, woods and fences, obstacles, demolitions, intrenchments, and other works.

When made of a position occupied by, or near an enemy, it will necessarily be hurried. But for defence there is usually sufficient time, and although it is not the duty of the reconnoitring officers on service to offer suggestions as to the best way of occupying a position,

Principles.
unless specially ordered to do so, certain principles should be borne in mind, viz.—there should be depth to the rear for the circulation of troops, the position should not be too extensive for the force to occupy it, and the flanks if possible secured, with safe and easy communication between them, and suitable ground for manœuvring in rear. The greatest number of

Plate X

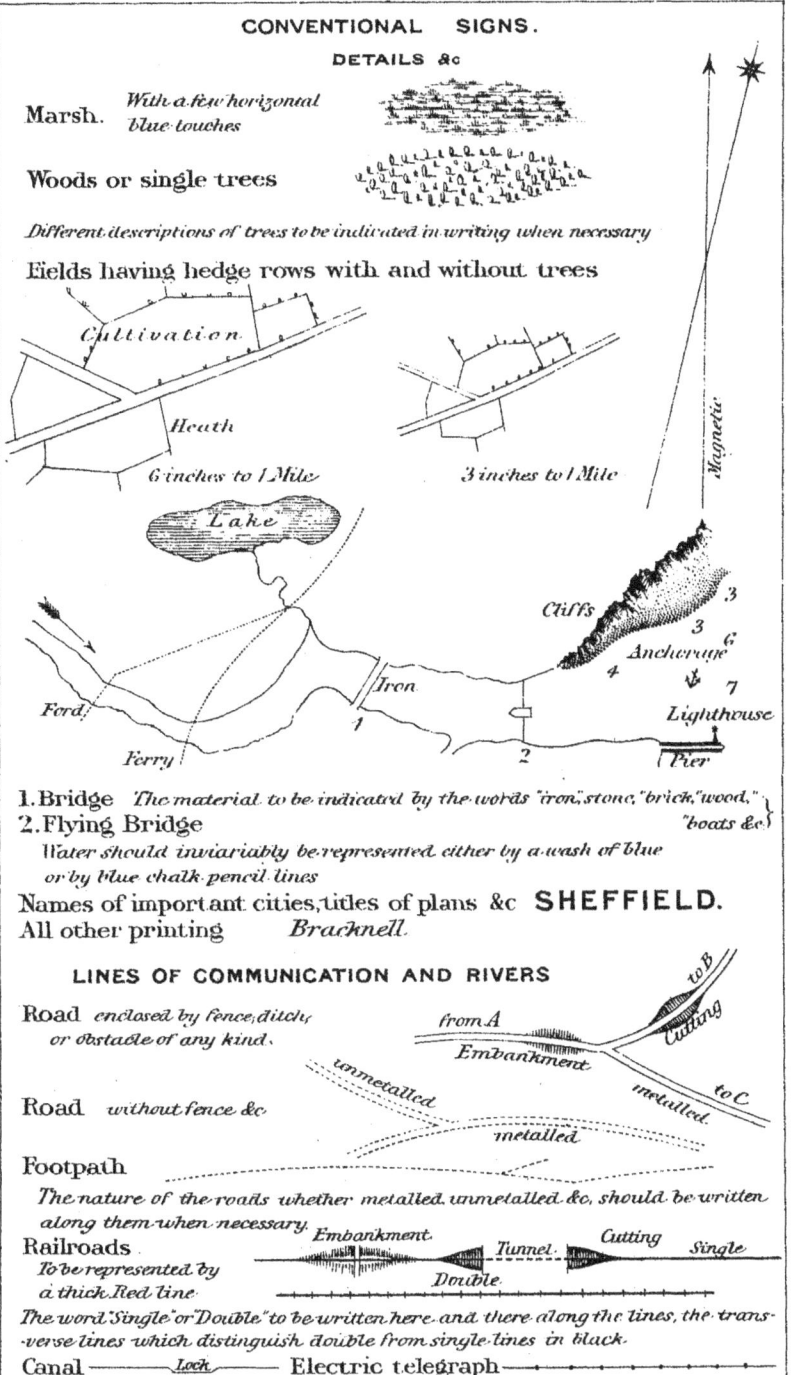

CONVENTIONAL SIGNS.

Fence, bank or Wall

Nullahs (India)

Church or Chapel

Other buildings } and gardens

6 inches to 1 Mile   3 inches to 1 Mile

THE FOLLOWING TINTS SHOULD BE USED WHEN POSSIBLE

Woods	Light shade of green (Hooker's N° 2.)
Cultivation	Light shade of gamboge
Main roads	Burnt sienna
Masonry	Crimson lake
Wooden building	Black
British troops	Crimson lake
Opposing force	Blue

When the same troops are shewn in successive positions, different shades of the same color should be used.

For rough sketches coloured chalk pencils (blue, brown, red and green) may be used.

TROOPS &C.

Scale 6 inches to 1 Mile.   3 inches to 1 Mile.

Field Artillery   Battery

Infantry { 500 men in 6 Co.s .... in column
           two deep ▭ in line

Cavalry { 500 men in 4 ▯ ▯ ▯ ▯ in column
          squadrons ▭ ▭ ▭ ▭ in line

Picquet   □
Support   □   Direction of a Patrol ➤
Sentry   ○   Double Sentry   ○ ○
Vedette   ○   Double Vedette   ○ ○

Day Sentries, Vedettes, and Patrols, coloured red.
Night do   do   do   uncoloured

Baggage   ⌂

Clearances or demolitions

Entanglements or abattis

Intrenchments

Rifle pits

Battery

Site of battle

Scale.
100 0   200   400   600   800 Yds

Scale 6 inches to a mile R.F. $\frac{1}{10560}$ Contours at — ft. vert.l int.l or
100 0   200   400   600   800 Yds.

Plate X. cont.d

troops should be devoted for the defence of vital points. The position should command a good view to the front, with gentle slopes for effective action of artillery. The requisites for a good position for that arm have already been noticed in detail (*vide* p. 33). A reserve must be in hand to support any part attacked. The three arms should be distributed on the ground best suited to their action, and certain points should be chosen well to the front as advanced posts.

Retreat should be easy, and another position selected to be taken up if necessary a little further to the rear, also arrangements made for pursuit in the event of success.

A danger has to be guarded against of extending too far in order to seize a vital point, and of occupying too much ground for the available troops.

It is better to keep a small force concentrated and well under observation.

There are no rules as to the strength to occupy a certain extent of ground in line of battle. Two examples may be quoted: At Alma the Russians occupied 4 miles with 40,000 men and 110 guns. which gave about 12,000 men and 36 guns per mile. At Waterloo, the British, with their Dutch-Belgian allies, numbering 75,000, held a front of 2 miles. *Strength per mile. Alma. Waterloo.*

The points to notice on a canal resemble those relating to a river, and include the position and dimensions of locks and what means they afford of crossing. The space between two locks may be run dry by opening the gates of the lower lock. *Canal.*

The places between which the canal communicates,

the tonnage of the craft navigating it and the condition of the tow-path, should be stated.

Marshes. * With regard to morasses and marshes we should observe their extent, and if it is possible to pass them by any paths, or if not to make a corduroy road of rough felled timbers or by laying faggots of brushood, furze, &c.

Ice. * Small parties of infantry can cross ice 3 inches thick, cavalry and field-guns 4 to 8 inches thick. In 1881 rails were laid across the St. Lawrence at Montreal, and trains ran over the ice for some weeks in winter.

Questions to inhabitants. * Inhabitants should be questioned as to the population, the name of the mayor and principal persons at a certain place, the number of horses and vehicles, the strength and position of the enemy's troops, his outposts and patrols. It is well to note the answers given to use when questioning others.

Questions to prisoners. * Prisoners should be questioned as to their battalions or corps, the names of their commanders, the position of other corps, the places where they marched from, the condition of the soldiers, if they are well fed, the position of any intrenchments.

Observations of the enemy. * A reconnoitring party, observing an enemy on the march, may ascertain his numbers by noting the time occupied by the different arms in passing a certain point, their breadth of front, and rate of marching. Thus—if a spy reports that a column, consisting of 9 battalions (1,000 strong) in fours, preceded by a cavalry regiment (384 sabres) in sections, and followed by 2 batteries of field artillery in column of route, passed through a village, taking 47 minutes to pass his house, and that he observed the usual intervals between the corps, we can prove that his statement is accurate as under :—

Cavalry, 384, in sections	= 384 yards.
Infantry, in fours, $\dfrac{1,024}{3} \times 9$	= 3,072 ,,
9 intervals of 25 + 1 of 28½	= 253½ ,,
2 field batteries @ 224 + 1 battery interval 9½ =	457½ ,,
	4,167 yards.

At 3 miles an hour $\dfrac{4,167}{88}$ = 47 minutes.

In the Peninsular War a Captain Grant concealed himself in a tree, counted the entire French army passing under it, and reported its strength to the Duke of Wellington.

The simplest plan to estimate a force of artillery is to count the guns. But when this cannot be done, we may take all the guns and carriages to occupy 20 yards each on the march, and so allow for intervals.

We can recognise masses of troops at a mile distant, and tell infantry from cavalry, and see troops moving at 1,300 yards, but the cavalry horses are not distinct. A single man is seen at 1,100 yards, but his head is not a distinct ball till 700 yards. We can make out white cross-belts at 600 ; and the face as a light spot, the uniform, head, arms, &c., and their movements, at 500 yards; buttons at 250 yards, and officers are distinguishable from the men. Bright scabbards and fixed bayonets are visible on a clear day about 1,500 yards. * Indications of troops.

Judging distance by eye rapidly is of great assistance, especially when it is remembered that in * Judging distance.

making a road sketch, with sufficient information on either side, the rate of a mile an hour is thought good. On fine clear days large objects on high ground are visible at considerable distances—*e.g.* a tower or church spire about 10 miles; a windmill 5 or 6 miles, &c. Window panes can be made out about 500 yards.

Sound. * As sound travels about 350 yards a second, by multiplying the time in seconds between the flash of a gun and its report by 350 gives the distance of the gun. Thus, if the time is 5 minutes, the distance will be about a mile.

The voice is heard about 150 yards; musketry and a band of music on a still day 2 to 3 miles; heavy guns about 20 miles—*e.g.* naval practice at Portsmouth is frequently heard at Southampton.

Forage. * It is desirable to be a good judge of forage. We can recognise oats to be of good quality when they
Oats. * are clean, hard, dry, sweet, plump, full of flour, and rattle like shot; and they have a clean and almost metallic lustre. Those in a sample vary very little in size; they are entirely free from smell, and the pressure of the nail ought to leave little or no mark on them. The value depends mainly on the weight per bushel. Dirty oats will weigh heavier than clean.

There are four possible defects in oats, viz.—being kiln-dried, foxy, damp, or dirty. New oats are known from old by having a fresh, earthy smell, and a milky taste, while the taste of the latter is slightly bitter. Oats run from about 38 to 42 lbs. the bushel.

Best hay contains a large proportion of good *Hay.* grasses—*e.g.* rye grass, foxtail, timothy, sweet vernal, &c., together with clover, and other good herbage, and but a small proportion of bad grasses; while in inferior hay good herbage is almost entirely wanting, and bad, rank grasses predominate.

We can distinguish good hay by its green colour combined with a delicate taste and smell, a presence of flowers in their natural colours, a variety of grasses, good herbage (not weeds), and clover. It should be moderately fine, crisp, and hard; musty and newly-mown hay are bad. Green fodder is preferable to new hay.

An acre of good meadow yields 2 to 3 tons of hay, while an acre of poor land will scarcely yield 1 ton.

A truss of hay weighs 56 lbs., a truss of straw 36 lbs., a load of either hay or straw is 36 trusses. Old and well-pressed hay weighs nearly three times new hay.

Straw should be either wheat, oat, or rye. Barley *Straw.* induces skin diseases. Straw should be strong and long. Some horses are inclined to eat wheat straw. Rye straw is generally too expensive.

A daily ration for a horse is 10 lbs. oats, 12 lbs. *Rations.* hay; and in quarters 8 lbs. straw, with 7 gallons of water at least a day. If no oats are procurable he should have 32 lbs. hay instead.

An ordinary ox will give 300 rations of meat, a sheep 45, a pig 110.

The volume of a hayrick of ordinary shape is found to be equal to $\left(a + \dfrac{b}{3}\right) \times c \times d$ (*vide* Fig. 136).

<small>Contents of a rick of hay or straw (vide General Order 130 of 1880).</small> And the volume of the usual form of a rick of straw is found by treating it as a cylinder on the same base, and of the same height. Then if h is the height, and r the radius equal to half the base, roughly the volume $= \pi\, r\, h$.

Conclusion.

The foregoing pages comprise the most essential elements of field-training. But as there is no limit to knowledge, it rests with each individual, not only to retain what he has already acquired, but also to progress with the age, and so be ready at any time to prove his ability when occasion offers for active service in the field.

An officer will do well to exert himself fully in this training, and infuse his own personal energy among his subordinates, exacting from them evidence of capacity in their respective positions. Some he will find inferior in power of thought and enterprise to others. But in spite of difficulties, he should induce the belief that his men belong to the *best* regiment in the Service, and compose the *best* company in that regiment. Thus, all being bound together by the same *esprit-de-corps*, the weak points of the few are merged in the efficiency of the majority, and success is certain.

INDEX.

A.

ABATTIS, 162
  in defence of a wood, 146
Accommodation, 240
  to be tabulated, 242
Acre, yield of crops in, 242
  yield of hay in, 249
Action, conduct of soldier after, 66
  conduct of soldier in, 56, 62
  employment of artillery in, 30
  employment of cavalry in, 24
  employment of infantry in, 45
  no artillery reserve in, 42
  three modes of cavalry, 25
  two modes of infantry, 43
Adjusting sights, 59, 64
  at close range, 58
Advance across bridge in presence of enemy, 55
  of infantry, 50
  to attack, 4
Advanced guard, composition and strength of, 70
  distance in front of army, 75
  division of, 71
  duties of commander of, 79
  duties of commander of support of, 79
  entering a defile, 75
  forms the outposts, 86
  general duties of, 70
  must protect main body at all risks, 80
  necessity for, 69
  of a battalion on a road, on a plain, 73
  of a cavalry regiment, 74
  of infantry in mountains, 70
  of the three arms, 74
  on first sighting the enemy, 80
  principle of formation of, 72
  proportion of, to army, 70
  reason for cavalry with, 73
  should not pursue, but maintain touch, 80
  specially organised, 70
  support of, 73
  to be accompanied by signallers, 71
  to be furnished with axes and tools, 71
  when accompanied by guns, 74

Advanced guard, with and without cavalry screen, 70
Advanced post, village as, 141
Agents, moral, 2
Alarm post for piquet, 99
Alcantara, bridge over Tagus at, 218
Alert, outposts to remain on till all posted, 90
  the only bugle sound, 5
Allotment of ground to outposts, 88
Allowance of forage and rations, 249
  of fuel, 17
  of water in camp, 199
Alma, strength per mile at, 245
Alva, crossing of the, by Massena, 214
Ambulance follows regiment on the march, 112
Ammunition, artillery, 40
  carried by the soldier, 54, 62
  carts, 54
  replenishing, 54
Anchoring flying bridges, 212
Aneroid barometer, 236
Angle, excess of earth at salient, deficit at re-entering, 22
  forward, back, closing, check, 230
  salient, abattis at, 146
  to lay off backwards, 232
  to lay out a right, 10
Animals, rations for, 249
  time taken by, drinking, 241
Arch blocked for inundation, 167
Arms, how carried in action, 56
  how slung for escalade, 190
  how slung for shelter trench exercise, 18
  how stacked in camp, 12
  not to be placed on baggage waggons, 14
  of artillery, 31
  of cavalry, 25
  of infantry, 43
  the three, necessity for a knowledge of, 24
  the three, on outpost duty, 104
  the three, proportion of, 4
Arm racks extemporised, 12
Army corps, composition and organisation of, 5

Army Corps, strength of, 6
Arrangements against artillery fire in villages, 139
  for a convoy, 128
  for a march, 111
  for assault and escalade, 185
  for attack, 3
  for loading baggage waggons, 14
  for tent pitching, 12
  for water in camp, 8
  for water supply, 240
Articles for reconnoitring, 221
Artillery, attacking a bridge, 154
  attacking a ford, 155
  cover for, 35, 183
  crossing a river, 157
  duties in attack, 30
  duties in attack of a house, 137
  duties in attack of a village, 143
  duties in attack of a wood, 150
  duties in defence, 30
  duties in defence of a village, 140
  duties in defence of a wood, 147
  duties in defending a bridge, 153
  duties in flank march, 120
  duties in forcing a defile, 152
  duties of, 29
  duties with advanced guard, 74
  duties with outposts, 86, 104
  duties with rear guard, 83
  effective range of, 3, 38, 42
  escort, conveyed on waggons of, 68
  escort, duty of officer commanding, 67
  escort, formation on the march of, 67, 68
  escort, how applied for, 67
  escort, necessity for, 67
  fire, effect of, 37, 88
  fire, object of, 30
  helpless in motion, 31
  in action, 3
  limit of strength in, 41
  organisation of, 40
  passing through a line, 41
  position, 6, 33
  power of, 31
  projectiles, 37
  proportion of, 41
  rate of marching, 117
  space occupied by, in line, 32
  space occupied by, on the march, 116
  tactical unit, 32
  various kinds of fire, 36
Assault, conduct of, 62, 185
  time for, 185
Assaulting column, division of, 186
Attack, arrangements for, 3
  artillery in, 30
  broken ground best for, 7
  conduct of piquet against, 103
  false to be distinguished from real, 130
  formations of cavalry for, 28
  hour for, 102
  how to practise the, 53
  of a barricade, 160
  of a convoy, 133
  of a position, 49—53
  of a village, 143
  of a wood, 149
  on a house, 137
  on Le Bourget by the Germans, 144
  stages of infantry, 50

B.

BAD roads, effect of, 122
Baggage, arrival of, in camp, 11
  convoy of, 126
  guard, duties of, 14, 129
  laagering convoy of, 131
  light, by rail, 124
  loading pack animals with, 195
  on the march, 112
  parking convoy of, 127
  waggons, loading of, 14
Balaklava, the "Thin Red Line" at, 60
Barracks, interior economy in, xvi
Barrel bridge, 200
Barricade, attack on a, 160
  how improvised, 159
  of doors and windows, 136
Barrows save labour in field works, 182
Base of operations, 1
Bâteaux of foreign armies, 207
Battalion, advanced guard of, 73
  advancing across a bridge, 55
  arriving on camping ground, 10
  attack of a, 50
  camp of, 9
  front of, in line, 44
  flanking parties of, 81
  in retreat, 55
  length in column of fours, 115
  rear guard of, 82
  re-forming, 54
  retiring across a bridge, 56
  skirmishing, 46
  squares, 62
  transport by rail of, 124
  war establishment of, 45
Battery, camp of, 14
  front of, 32
  length in column of route of, 116
  number of guns in, 32
  rarely broken up, 41
  tactical subdivision of, 40
  war establishment of, 40
Battle of Tel-el-Kebir, 121
  of Waterloo, 45, 245
Bayonet against men with shields, 48
  charge, 22, 53
  fixing and unfixing on the march, 54
  importance of, 43
Berm, 20, 187
Bickford's fuse, 193
Bivouac, 13
Blankets packed first in waggons, 14
Blowing in a gate, 194
Boarding up windows, 169
Boats, rowing, 212
Boil, time water takes to, 17
Bombproof barracks, 174
Book-field, 281
Branch roads, to explore, 77
Breakfast before starting on the march, 114
Breastworks, different forms of, 172
  in American Civil War, 147, 173
Bricks, 243
  houses of, better than stone for defence, 135
Bridges, American sheet-iron waggon, 209
  anchoring flying, 213
  attack on, 154
  best position for, 153

Bridges, collecting materials for, 157
cribwork, 210
defence of, 153
different kinds of, 199
examination of, 77
flying, 212
frame, 217
iron gabion, 217
ladder, 216
lever, 215
not to be broken without orders, 100
of Belgian trestles, 208, 211
of boats, 207
of rope at Alcantara over Tagus, 218
of wooden trestles, 210
on line of march, 239
over Adour, 208; over Prah, 210; over Beresina, 210; over Coa, 215
picket at, 100
pontoon, 206
protection of, 208
raft, 209
separate for different arms, 56
suspension, 217
trail, 213
weights on, 206
Brigade, composition of cavalry, 5
composition of infantry, 4
disposition of, on outpost duty, 105
Bugle sounds disclose intentions, 49
the only, allowed, the "Alert," 57
Bullet, penetration of, 21

C.

CABLES for flying bridges, 213
in bridge over Tagus at Alcantara, 218
Calculation at Chatham for cutting brushwood, 175
of discharge of a small stream, 199
of volume of rick of hay or straw, 249
Camp, arriving in, 10
colourmen, 9
formation of, 11
internal arrangements of, 10
laying out a, 9
of a battalion, 9
of artillery, of cavalry, 14
reports on ground for, 241
sanitary arrangements in, 9
selection of a site for a, 8
striking, 12
Canal, reconnaissance of, 79, 245
Cap, peak of, to measure stream with, 219
Carriages, least width to pass for, 32
railway, capacity of, 124
Carts, ammunition, 54
issuing tools from, 19
Case, never fired over other troops, 39
range of, 39
rounds per gun, 40
when employed, 39
Casks, bridge of, drill to make, 201
piers of, 200
Cattle, rations furnished by, 249
Cavalry, action of, 25
camp, 14
charges, 28
chiefly used to attack a convoy, 138
columns, depth of, 27
distance at which distinguishable, 247

Cavalry, escort, disposal of, 68
flanks of vulnerable, 29
frontage of, 26
in advanced-guard, 73
in defence of a wood, 148
infantry opening for, 58
patrol, 77
piquet, 108
rate of march of, 117
regiment, advanced guard of, 74
scouts, 29
screen, 24, 70
space occupied by, 26
space on the march of, 27
speed of, 26
tactical unit of, 26
use of, 24, 29
war establishment of regiment, 28
when in support on outpost duty, 109
Cease firing, signal for, 57
Cellars as protection, 135
Charge from trenches, 22
of cavalry, 25
of infantry, 64
to capture guns, 29
Charger pit, 22
Chatham calculation of brushwood, 175
Chesses, 205
Chevaux-de-frise, 166
Circle, graduation of compass card, 227
Cities, accommodation in, 240
reconnaissance of, 240
Clearances, 243
Clear field of fire, 135, 147
Clinometer, how made, 234
Code of signals, 71, 94, 106
Column of half-batteries and divisions, 36
of infantry, cavalry, artillery, 115
of mixed force, 117
of route, meaning of, 115
of squadrons, of troops, 27
opening out, bad effects of, 112
space occupied by, 44, 45
Combat, single, 60
Commander of advanced guard, 79
of escort, duties of, 67
of outposts, 88
of piquet, 89, 97
of rear guard, 82
of vanguard, 71
section duties of, 48
Commanding officer's dispositions for attack, 50
Command over enemy's position, 34, 139
Common shell, 37
Communications, line of, 2
marking, in woods, 146
none through keep, 143
none through tête-de-pont, 153
over parapet, 173
through walls, 136
Company, directing, 48
front of, extended, 44
officers, how to train their men, xiv, 250
squares, 61
the fighting unit, 43
war establishment of, 45
Compass, errors of, 228
measurements by, 228
prismatic, 227
use of, 228
variation of, 230

Compliments, none from outposts, 105
Conduct of assault, 62
　of convoy, 126
　of march, 110
　of piquet when attacked, 103
　of soldier in action, 56, 62
Connecting files, 72, 92, 114, 132, 146
Construction of charger pit, 22
　of embrasures, 183
　of field kitchens, 15
　of hasty field-work, 183
　of loopholes, 169
　of obstacles and revetments, 161, 174
　of rifle pits, 184
　of shelter pits, 19
　of shelter trenches, 20
　of single lock bridge, 216
Contents of hay or straw rick, 249
Contours, 234
Conventional signs, 244
Convergence of roads, best place for, 101
Convoy, attack of, 133
　conduct of, 126
　escort of, 127
　of prisoners, 128
　rate of march of, 129
Cooking in the field, 15
　on piquet, 102
　party in camp, 11
　rules for, 15
Coolness in action, 59, 63
Corn, 248
Corps, army, 6
Counter attack, 45, 55
Counting distances, 233
Cover, artificial, 18, 172, 173
　denied to enemy, 83
　for guns, 35, 183
　from view, 6
　natural, 7, 50
Cribwork, 210
Crops, information regarding, 242
Crossing rivers, 198
　selection of point for, 157
　the Alva, 214
　the Beresina, 210
　the Douro, 158
Crowsfeet, obstacles, 165

D.

Dam, for inundation, 166
Daybreak, hour for relief of outposts, 102
　usual hour for attack, 102
Defence, artillery in, 30
　flank, 99, 136, 167
　infantry in, 65
　of a bridge, 153
　of a ford, 154
　of a position, 244
　of a wood, 145
　of houses, 135
　of posts, 161
　of rivers, 156
　of villages, 138
　open ground in front, best for, 7
　works auxiliary to, 161
Defile, advanced guard entering a, 75
　convoy passing a, 129
　meaning of a, 150

Defile, mountain, 152
　passing on the march, 113
　rear guard holding a, 83, 152
　retreat through a, 152
Defiling, delay resulting from, 113
　forbidden, unless by preceding troops, 113
Definitions, 1
Demolitions, hasty, 191
　of a bridge, 191, 194
　of a gate, 193, 194
　of a stockade, 192
　of a railway, 194
　of guns, 195
　of houses, 194
　of palisades, 192
　of telegraphs, 195
　of trees, 194
　of walls, 191
Deployment, necessity for, 4
　time required for, 118
Depth for, fords, 154
　for positions, 244
　of cavalry columns, 27
　of infantry columns, 44
Deserter, from piquet, precautions to be taken in case of, 95
　how to be received by outposts, 95
Detached duties of cavalry, 24
　post, house as, 135
　post, village as, 141
　post, when necessary, 96
Detonator, Bickford's fuse to be well fixed in, when firing charges, 193
Detraining infantry, 125
Diggers, best to be in rear rank for shelter trenches, 22
　proportion of, to rammers and shovellers, 181
Directing company or file, 48
Discharge of a small stream, how to find, 199
Discipline, danger in relaxing, xiv.
　strict, essential in "extended order," 46
　to be enforced on the line of march, 112
Disembarkation from a train, 125
Disks of guncotton, 193
Dismantling a bridge of casks, 206
Distance between artillery and other arms, 116
　between battalions, 116
　between cavalry and infantry, 28
　judging, 57, 247
　measured by sound, 248
　measured by time as well as length, 227
　measurements of, 233
　of advanced guard in front of an army, 75
　of outposts in front of the main body, 87
　of rear guard from an army, 81
Distribution of battalions into brigades and divisions, 5
　of working parties, 181
Ditch, crossing the, in escalade, 190
　cut outside a lower door or window in defending a house, 136
　object of steps in excavating the, 181
Division, as a unit, 4
　of guns, 32
　order of march of an infantry, 121
　personnel of, 5
　rate of march of a, 117

Division, the Light, under General Crawfurd, 123
Doors, barricading, 136
Doubling to be practised in running drill, xvi
   to be steady in "extended order," 51
Douro, passage of the, 158
Drains cut in camps, 9
Drill, escalading, 188
   for making bridge of casks, 201
   not to be deviated from without authority, xiv
   object of, xiv
   recruits, distribution of time for, xv
   running, object of, xvi
   shelter trench, 18
Drinking, time taken by horses, 241
Drivers, artillery, 40
   to be watched in convoy, 129
Dynamite, 191
   experiment at Chatham with, 192

E.

EARTH, how far thrown by workmen, 181
   more required at salients than re-entering angles, 22
   not to be heaped outside walls as obstacles, 171
Echelon, flank battalion attacking to be in, 52
   squares to move into, 61
Effective range of common shell, 38
   of case, 39
   of shrapnel, 38
Effects of a determined charge, 64
   of artillery and infantry fire, 3
   of bad roads, 122
   of roads blocked with baggage, 122
   of tailing off on the march, 112
Embrasures, construction of, 183
   infantry in a siege to fire at, 62
Employment of artillery in action, 30, 42
   of cavalry in action, 24
   of infantry in action, 42
   of shrapnel in Egypt, 39
Encampments, 8, 241
Enemy, observations of the, 246
   position of, to be reported on, 78
   skeleton to be used for exercise, 53
   to be deceived, 209
   to be kept out of a wood, 145
   to be stalked in the attack, 57
   touch of, to be maintained, 78, 80
   when first sighted by advanced guard, 80
Engineers, duties of, 186
   necessity for, 43, 142
   to accompany advanced guard, 73
   to accompany rear guard, 82
   with storming party, 144
Enlarging a map, 227
Entanglements, wire, 140
   wood, 163
Entraining infantry battalion, 125
Entrenching tools, how carried, 18
   how issued from heaps or a cart, 19
   how laid out, 18
Entrenchments, 18, 20, 35
   in villages, 140, 141
   in woods, 147
Epaulments for guns in a village, 140
   for guns in a wood, 147

Escalading, 185
   drill, 188
Escarpment, 167
Escorts for artillery, 67
   to foraging parties, 134
   for prisoners, 128
Esprit-de-corps, 250
Establishment, war, of a battalion, 45
   war, of a cavalry regiment, 28
   war, of a company, 45
   war, of a squadron, 28
   war, of batteries, 40
Examination by patrol of a bridge, 77
   by patrol of a farm, 76
   by patrol of a wood, 76, 146
   of inhabitants and prisoners, 246
Examining party, 95
Example, advanced guard of brigade, 74
   advanced guard of small mixed force, 74
   breastworks, 147, 172
   brigade on outpost duty, 105
   exploring patrols, 107
   night march, 121
   notable marches, 122
   of distribution of a piquet, 91
   order of march of a division, 121
   simple scales, 223
   small force in column of route, 117
   time marches, 118
   use of hedges, 170
Expenditure of ammunition to be controlled, 49
Exploring a branch road, 77
   patrols, 107
Extended line in attack on a house, 137
   line in defence of a village, 139
   order distinguished from skirmishing, 46
   order, object of, 46
Extension of battalion for attack, 50
   of covering party, 19, 186
   of polemen, 11
   of supports, 51
   of working party, 20, 182

F.

FALLING out unnecessary on the march, 112
False attack to be distinguished from real, 130
Farmhouses, reconnaissance of, 76
Fascines, 177
   picketing in revetment, 178
Feeding horses of cavalry piquet, 108
Feet, treatment of sore, 114
Ferries, 213, 241
Field battery, column of route of, 115
   battery, rate of march of, 117
   book, 231
   cooking, 15
   fortification, accessory to defence, 161
   guns, cover for, 35, 140, 147
   guns, epaulments for, 140, 147
   guns, how destroyed, 195
   guns, how made unserviceable, 59
   guns, limit of opening for, 32, 239
   guns, passing through a line, 41
   guns with advanced guard, 74
   guns with outposts, 104
   guns with rear guard, 83
   officer of the day on outpost duty, 88

Field of fire to be cleared, 135, 147
  sketching, 228
  telegraph with outposts, 86
  training, xiv, 111, 250
  works, trace of, 180
  works, construction of, 181
Fighting line, 50
Files to reduce to paces, 44
Filling in trenches, 23
Finger posts, 143, 146
Fire, artillery and infantry, 3
  cease, signal for, 57
  precautions against, 137
  three zones of, 3
  time to be allowed for artillery, 143
  various kinds of artillery, 36
  various kinds of infantry, 57
  with outposts, 98
  wood ration of, 17
Firing, in attack, 52
  on the defensive, 65
  three military positions for, 58
Fitting shoes, 114
Fixing and unfixing bayonets on the move, 54
Flags of truce, 95
  signal parties with, 95
Flank defence, 136
  march, meaning of, 120
Flanks, infantry on, in mountains, 79
  protection of, 29, 89, 100
  to be turned, 77, 107
  troops on, to be in échelon, 52
Flanking parties, duties of, 75, 80
  to be vigilant with rear guard, 82
Flying bridges, 212
Forage, 248
  rations of, 249
  to, a village, 134
Foraging parties, escorts to, 134
Ford, attack of, 155
  defence of, 154
  depth for, 154
  piquet at, 100
  reconnaissance of, 154, 198, 239
Forest fighting, 145
  reconnaissance of, 146, 149, 241
  roads through, resemble defiles, 145
Form of reconnaissance, sketch and report, 237
Formation, defensive, 54
  for attack, 50
  of advanced guard, 72
  of artillery with other troops, 41
  of camps, 8
  of cavalry, 26, 28
  of escort for artillery, 67
  of rear guard, 82
  of working parties, 180
  to resist cavalry, 60
Forming bridge of casks, 205
Forward angle, 230
Fougasses, 167
Fours, column of, in cavalry, 27
  length of battalion in, 115
  use of, in rough ground, 52
Fraises, 165
Frame bridges, 216
Frederick the Great's maxim as to advanced guards, 79
French bâteaux, 207
  marches before Leipsic, 121

French neglect of outposts, 85, 159
Frontage of troops in position, 26, 32, 44
Fuel, party to collect, 11

G.

GABION, 175
  iron band trip, 166
  iron bridge, 217
  Jones's iron, 176
  mode of carrying, 177
Gallop, rate of, 26
Garrison of farm or group of houses, 137
  of house, 135
  of village, 142
  of wood, telling off of, 148
Gates, blowing in, 193, 194
Gatling gun, 140, 143
German advanced guards in 1870, 71
  attack on Le Bourget, 144
  instructions to infantry in action, 62
  proportion per mile on outpost duty in 1870, 87
Glass, remove from windows in defence of houses, 137
Gradients, practicable, 237
  practicable, of roads, 238
Grain, inspection of, 248
Ground, fitting shelter trenches to, 20
  for camping, 9
  in relation to tactics, 6
  most suitable to each arm, 6
  piquet marching to, 89
  soldiers to learn use of, 57
  working party marching to, 19
Groups, 61
Guard, advanced, 69
  baggage, 14
  examining, 95
  house in village, 143
  over prisoners, 128
  rear, 81
Guide to measurements in shelter trenches, 20
Guides, 120
Guncotton, 193
Gunners, 40
Gunpit, 188
Guns, ammunition for, 40
  cover for, 35
  in attack of a ford, 156
  in attack of a house, 137
  in attack of a wood, 150
  in defence of a village, 140
  in defence of a wood, 147
  in passage of a river, 157
  not to be in front or in rear of other troops, 35
  number per battery, 32
  passing through a line, 41
  proportion of, 41
  range of field, 3, 38, 39, 42
  with advanced guards, 74
  with outposts, 104
  with rear guards, 83

H.

HALF-BATTERIES, column of, 36
  sections of cavalry, 28
Halts on the march, 115, 129

Hasty conclusions to be avoided, 78
demolitions, 191
field work, construction of, 183
sketch, how to make, 222
Hay, inspection of, 249
rick, volume of, 249
yield of, per acre, 249
Heat, effect of, in marches, 112, 115
Hedges, treatment of, 171
Heights, relative, to be marked, 236
Helplessness of artillery in motion, 34
Horizontal equivalent, meaning of, 235
Horse, allowance of water to, 199
artillery, battery front of, 32
artillery, column of route, 116
artillery employed with rear guard, 83
artillery, pace of, 117
drinking, time required for, 241
length in ranks of, 27
rations for, 249
watering, 240
Hospital, how known in the field, 66
Hot meals when possible on outposts, 102
Hougomont, value of hedges at, 172
Hour for attack, 99, 102
Houses, accommodation in, 240
attack of, 137
defence of, 185
reconnaissance of, 76
Hurdles as revetments, 178

I.

Ice, bearing power of, 246
Improving banks, 173
Indications of troops, 247
Individuality of the soldier, 46, 48, 64
Infantry, advanced guard of, 73
ammunition, 54
arms, 43
as working parties, 181
attack, 49, 144, 150, 152, 154
brigade, 5
camp, 9
charges, 53, 64
crossing bridges, 56
deployment of, 4
distances at which distinguishable, 24
division, 4, 121
entraining, 124
escorting a convoy, 127
escorting artillery, 67
establishment in war, 45
fighting line, 50
fire, 3, 57, 58
flanking parties of, 70, 75, 80, 83
functions of, 6, 42
in action, 56, 62
in assault, 185
in a siege, 62
in defence, 45, 65, 140, 145
in extended order, 46, 140, 145
mounted, 25
on outpost duty, 84
on the march, length of, 111
organisation of, 4, 43
pace of, 43, 116
rear guard of, 82
receiving cavalry in line, 60
scouts, 49

Infantry, single combat of, 60
skirmishing, 46
space occupied by, 44
squares, 60, 62
tactical unit of, 43
transport of, by rail, 124
versus artillery, 59
versus cavalry, 59
versus infantry, 58
Information, nature of. *Vide* Reconnaissance.
tabulated, 242
to be obtained, 128, 149, 185
to be transmitted, 107
Inhabitants, agricultural, useful for intrenchments, 240
questions to, 246
Instruction of German infantry soldier in action, 62
of recruit, xv
to patrols, 78
Instruments for surveying, 221, 227
Interior economy in barracks, xvi
Interpolation, 229
Interrogating inhabitants, 246
prisoners, 246
sentries, 92
Intervals between artillery and other arms, 33
between cavalry and infantry, 28
necessity for, 41
Intrenching tools, how carried, 19
issue of, 19
laying out of, 18
Intrenchments, 18
Inundations, 166
Iron gabion bridge, 217
gabions, 176
girder bridge, destruction of, 194
Issue of tools from heaps or a cart, 19

J.

Jones's iron gabions, 176, 217
Judging distances, 57, 247

K.

Kettles, camp, 15
Kinds of bridges, 199
of cooking trenches, 15
of fire, 36, 57
of obstacles, 162
of patrols, 105
of revetments, 174
Kitchens for field cooking, 15
Knowledge of ground essential, 6, 57
of country essential, 220

L.

Laager to resist attack, 131
Labour in excavating earth, 172
Labourers, agricultural people as, 240
Ladders, bridge of, 216
for escalade, 188
in defence of a house, 136

s

Ladders, party for in, escalade, 187
Landmarks to be fixed in a night march, 120
Laths, profiling, 180
Latrines, 13
  party to make, 11
Laying off an angle backwards, 232
  out a camp, 9
Le Bourget, German attack on, 144
Length, horse's, 27
  of artillery teams, 32
  of columns on the march, 115, 116, 121
Lengthening bridge by trestles, 212
Lever bridge, 215
Light baggage at railway station, 124
Limbers and artillery waggons to be sheltered, 34
Limit to depth for fords, 154
  to escalade, 185
  to strength in artillery, 41
Line, fighting, 50
  front of troops in, 26, 32, 44, 117
  of communications, 2
  of observation, 88
  of operations, 1
  of resistance, 89
Loading baggage waggons, 14
  pack animals, 195
Load on bridges, 206
Lock bridge, 216
  on canal, 245
Logs on the top of walls, 170
Loopholes, dimensions of, 169
  for defence of barricades, 160
  in houses, 136
  of sand-bags and sods, 170
  on the tops of walls, 170

M.

MACHINE guns, 140
Magnetic meridian, 229
Manœuvres, gradients admitting of, 237
  to represent reality, xiv
Map, enlarging or reducing a, 227
  scales for a, 222
Marauding forbidden, 66
March, arrangements for the, 111
  attack when on the, 80
  baggage on the line of, 112
  discipline on the, 112
  flank, 120
  forced, 121
  halts on the, 115
  hour of starting on the, 114
  length of a, 121
  night, 120
  opening out on the, 112
  order of, for a division, 121
  punctuality on the, 118
  space occupied on the, 115, 116
  time occupied on the, 118
  to attack, 4
  weight to be reduced on the, 111
Marches, 110
  instances of notable, 121
  time, examples of, 118
Marching a battalion into camp, 10
  of a piquet to its ground, 89
  rate of, 116
  rate of, how influenced, 114

Marching, route, to be practised, 111, 239
  to the ground to entrench, 19
Marking communications, 146
Marshes, report on, 246
Materials for hasty demolitions, 191
  collected for bridges, 157
  strength of, 219
Maxims of Frederick the Great, 79
  of Napoleon, 2
Measuring distances, 233
  distances by sound, 248
  distances by time, 217
  distances by triangulation, 228
  guide to, in shelter trenches, 20
  width of stream by peak of cap, 219
Meat, how to cook, 17
Military, bridging, 199
  considerations prevail near the enemy, 110
  indications, 247
  pits, 165
  spirit and discipline, xiv, 112, 250
Moral agents, 2
Mountains, defile in, 152
  infantry employed in, 70
Mounted infantry, 25
  orderlies with infantry outposts, 86
  orderlies with strong patrol, 108
Movement of battalion by rail, 124
Mules loading with Otago pack-saddle, 195
Music, band of, how far heard, 248
Musketry fire, 3, 57
  proficiency in, how stimulated, xv

N.

NECESSITY for advanced guard, 69
  for deployment, 4
  for developing warlike character, 48
  for intrenchments, 18
  for escorts to artillery, 67
  for intervals, 41
  for knowledge of country, 220
  for outposts, 84
Night, changes for, made by day on outpost duty, 94, 102
  disposition of piquets by, 102
  march, 120
  precaution against surprise at, 93, 103
North point, how drawn on a sketch, 230
  variation of magnetic, 230
Notice to station-master of troops moving, 124
Number of casks to make a pier, 200
  of paces in so many files, 44
  of rounds per gun, 40
  of telegraph poles per mile to count, 233
  of troops, to estimate, 246
  of men per mile on outpost duty, 87
  of men per mile in positions, 245
  piquet to have a, 98

O.

OATS, 248
Observation, line of, 88
Observations of the enemy, 246
Obstacles, 161

Obstacles, abattis, 162
 at salients, 146
 chevaux-de-frise, 166
 conditions to be fulfilled by, 162
 crowsfeet, 165
 engineers and pioneers to remove, 43, 73, 186
 entanglements, 163
 escarpment, 167
 fougasses, 167
 fraises, 165
 gabion band trip, 166
 inundations, 166
 military pits, 166
 palisades, 164
 to guard against surprise at night, 161
 use of, 161
 value in positions, 161
Occupation of a village, 138
 of a wood, 145
 of ground by outposts, 88
Offensive to be taken when opportunity occurs, 45, 141
Officer, company, in camp, 11
 field, of the day, 88
 mounted, to conduct battalion to camp, 10
 on outpost duty, 89, 96, 97, 103
Offset in sketching, 231
Open ground in front best for defence, 7
Open hole trench, 17
Opening in obstacles, 162
 of infantry to let cavalry and guns through, 53
 out on the march, 112
Operations, base of, 1
 line of, 1
Order of battle, 4
 of march, 110
 of march of a division, 121
Orderlies mounted with outposts, 86
 with strong patrol, 108
Organisation of a corps, 5
 of a division, 4
 of artillery, 32
 of cavalry, 26
 of infantry, 43
Outposts, artillery with, 86, 104
 changes for night made by day, 102
 component parts of, 88
 composed of cavalry and infantry, 86
 cordon system and patrols in, 85
 detached post, when necessary in, 96
 distance in front of main body of, 87
 duties and disposition of, 85
 flanks to be protected by, 89
 formed of advanced guard or freshest troops, 86
 how to practise, 109
 instances of neglect of, 84
 meaning of, 84
 mounted orderlies with, 86
 no compliments paid by, 105
 no hard and fast rules for, 98
 not to shut themselves up, 100
 of a brigade, 105
 principle of detailing troops for, 87
 proportion of, to army, 87
 relief of, 102
 remain alert till all are posted, 90
 sentries of, 91
 signals arranged for, 94

Outposts, signal stations with, 86
 to be under arms before daylight, 102
Ox, rations furnished by an, 249

P.

PACE of cavalry, 26
 of convoy, 129
 of field artillery, 32
 of horse artillery, 32
 of infantry, 43
 of large bodies of troops, 117
Paces, to reduce files to, 44
Pacing, 233
Pack animals, loading of, 195
Pahs, New Zealand, 192
Palisades as obstacles, 164
 destruction of, 192
Parade, never, sooner than necessary for a march, 110
Parallels made by infantry, 177, 182
Parapet of breastworks, 172, 173
 of parallels, 177, 182
 of shelter trench, 21
Parking convoy, 131
Parties detailed on arriving in camp, 11
 working, 144, 180
Passage of Beresina, 210
 of Bidassoa, 159
 of Danube, 209
 of Douro, 158
 of rivers, 157
Paths in woods, 146
Patrols ascending a hill, 77, 107
 examining farm, 76
 examining wood, 76
 exploring, 106
 exploring a branch road, 77
 instructions to, 78
 object of, 85, 106
 of cavalry, 77
 strong, 108
 to turn flanks, 77, 107
 visiting, 105
Penetration of bullet, 21
Petersburg, bombproof barracks at, 174
Pickaxe in shelter trenches, 20
 used for blowing in a gate, 194
Pickets as obstacles, 165
 fascine, 178
Piers of casks, 200
 of cribwork, 210
 to form, 200
 to launch, 203
Pig, rations furnished by, 249
Piles, 218
Pioneers to clear obstacles, 43, 73, 176
Piquets, cavalry, 108
 may not generally light fires, 98
 numbered, 98
 position of, 99, 102
 relief of, 102
 report, 98
 to afford mutual aid, 99
 when attacked, 103
 when to be under arms, 99
Pitching tents, 12
Pits, charger, 22
 gun, 35, 183
 military, 165
 rifle, 19, 184

Pits, shelter, 19
Plain, advanced guard crossin , 74
Planks as revetment, 178
Plans, how made, 221
   scales for, 223
Plevna, bombproof barracks at, 174
   defenders outnumbered at, 46
   effect of shell at, 38
Plotting, 232
Point for crossing a river, 157
   north, 230
   of advanced guard, 72
Pontoons, boat, 207
   cylindrical, bridge of, 206
Population, estimate of, 240
Position for bridge over a winding river, 153
   observations on enemy's, 78
   occupation of a, 244
   of piquets, 99
   of reserve, 104
   of sentries on outpost duty, 94
   of supports, 101
   reconnaissance of, 242, 244
   to hold a defile, 151
Post, detached, 96
   to be strengthened, 100
   of honour, command of rear guard, 82
Potatoes, how to cook, 17
   yield per acre, 242
Powder compared to guncotton, 191
Power of artillery, 31
Practicable slopes, 237
Practice in marching, 111, 118
   in outpost duty, 109
Precautions against fire in defence of a house, 137
   against surprise of a convoy, 129
   against surprise of a piquet, 100
   against surprise of a sentry, 93
   in choosing a camping ground, 8
   when one of a piquet deserts, 95
   when fires are allowed on outpost duty, 98
Preliminary instruction of recruit, xv
Principles deciding distance between piquets and supports, 101
   of defence of villages, 139
   of detailing outposts, 87
   of employing artillery, 30
   of forming an advanced guard, 72
   of hill shading, 235
   of occupying a position, 244
Prismatic compass, measurements by, 228
   use of, 227
Prisoners, escorts for, 128
   questioning, 246
Profiling laths, to erect, 180
Projectiles, artillery, 37
Proportion between the three arms, 4
   of advanced guard to army, 70
   of guns to men, 4, 41
   of men to front line of defence, 140, 148
   of men to shelter trench, 20
   of outposts to army, 87
   of rear guard to army, 81
   of reserve of advanced guard, 71
   of reserve of battalion in attack, 51
   of reserve of outposts, 104
   of supports to fighting line, 51
   of supports to piquets, 101

Protection of bridges, 208
Protractor, use of, 232
Pursue, advanced guard not to, 80
   cavalry to, 24

Q.

QUARTER GUARD, mounted by field officer in camp, 11
   master in camp, 9
Questioning inhabitants, 246
   prisoners, 246
   sentries, 92
Quick time, rate of, 43, 116

R.

RACKS, arm, extemporised, 12
Rafts, bridge of, 209
Rails, to twist, 195
Railway, capacity of, carriages, 124
   destruction of, 194
   report on, 78, 242
   transport by, 124
Rammers, 181
Range, effective, of artillery, 3
   of case, 39
   of common shell, 37
   of musketry, 3, 57
   of shrapnel, 38
   to be taken, 55
   to be wide for artillery, 33
Rate of marching, 26, 32, 116, 117, 129
Rations from pig, ox, sheep, 249
   of forage, 249
   of wood, 17
   party to draw, in camp, 11
Rear guard, duties of, 81
   how withdrawn, 83
Reconnaissance, articles for making, 221
   meaning of a, 220
   of bridges and fords, 239
   of camping ground, 241
   of canal, 245
   of country, 241
   of positions, 78, 244
   of railway, 242
   of river, 79, 241
   of road, 78, 238
   of town or village, 79, 240
   of wood, 79
Reducing a map, 227
Reforming battalion after attack, 54
   after taking a wood, 150
Regimental reserve ammunition, 54
Reinforcing by reserve, 52
   discretionary, 51
Relative heights to be shown, 236
Relays in working parties, 181
Relief of outposts, 103
   of sentries, 94
   to be kept separate, 99
Repair of roads, means for, 238
Replenishing pouches from ammunition cart, 54
Report, on bridges and fords, 239
   on canal, 245
   on camping ground, 241
   on positions, 78, 244
   on piquet, 98

Report, on railway, 78
 on reconnaissance, 237
 on river, 79, 241
 on road, 78, 238
 on town or village, 78, 24
 on various kinds of, 243
 on woods, 79
Representative fraction of a scale, 223
Reserve, advanced guard, 71
 ammunition, 54
 cavalry, 28
 employment of, 52, 55, 138, 148, 156
 in assaults, 138, 187
 necessity for, 52, 55, 138, 141, 148, 156
 none of artillery, 42
 of battalion attacking, 52
 of outposts, 87, 104
 reinforcing with, in attack, 52
Resistance, line of, with outposts, 89
Retiring battalion across a bridge, 56
Retreat, battalion in, 55
 cavalry in, 25
 facility for artillery, 34
 of beaten army, 81
 rear guard covers, 81
 through a defile, 151
Revetments, 174
 fascines as, 177
 gabions as, 175
 hurdles as, 178
 planks as, 178
 sand-bags as, 179
 sods as, 178
Rifle pits, 146, 184
 range of, 3, 57
River, defence of, 156
 passage of, 157
 reconnaissance of, 241
 velocity of, 198
Roads, length of columns on, 111, 115
 rail, report on, 78, 242
 reconnaissance of, 78, 238
 where, should converge, 101
Rope bridge over the Tagus, 218
Rorke's Drift, 161
Rounds carried by infantry soldier, 50, 62
 per gun of artillery, 40
Route, column of artillery, 115
 column of cavalry, 115
 column of infantry, 111
 marching, to be practised, 111
Rowing boats, 212

S.

SADDLE, Otago pack, 195
Salients, abattis at, 146
Sand-bags as revetments, 179
 loopholes of, 170
 to hold powder, 193
 to muffle ladders, 188
Sanitary consideration of camping ground, 8
Scale, plain, to construct, 222, 223
 representative fraction of, 223
 to be shown on sketch, 230
Scouts, cavalry, 29, 74, 78
 infantry, 49
Screen cavalry, 70
Section commander, duties of, 47, 48
 of cavalry, 27, 115
 in escalade, 188

Security of prisoners, 128
 rendered by outposts, 84
Selection of site for camp, 8
Sentries, connecting, 92
 double, how posted, 91, 92
 outposts not to take off valises, 93
 over arms, 97
 pay no compliments on outpost duty, 105
 position of piquet in regard to, 99
 posted in a wood, 94
 protected from surprise, 93
 questions to be answered by, 92, 94
 relief of, 94
 same to mount same posts, 93
 to be economised, 92
Sheep, rations furnished by, 249
Shell, common, 37
 shrapnel, 38
Shelter on enemy's side of river, 157
 pits, 19
 trenches, 18, 141
Shooting line, 139, 145
Shovellers, proportion of, 181, 182
Shovels in shelter trenches, 20
Shrapnel shell, 38
Siege, infantry in a, 62
Sights to be adjusted in action, 58, 59, 64
Signal code, 65, 71, 94, 106
 preferred to bugle sounds, 49
 stations with outposts, 86
 to cease firing, 57
Single combat, 60
Site for camp, selection of, 8
Skeleton enemy for instruction, 53
Sketching field, 221, 222, 235
 scales for, 222
Skirmishing, 46
Slabs of guncotton, 193
Slinging arms for shelter trench drill, 18
 for escalade, 190
Slopes, how represented, 233
 practicable for manœuvres, 237
 practicable on roads, 238
Small arm ammunition cart, 54
Socks, examination of, 114
Sods, loopholes of, 170
 revetment of, 178
Soldier in action, 56, 57, 62, 66
 weight carried by, to be reduced, 111
Sore feet, treatment of, 114
Sound, velocity of, 248
Space for troops in position, 26, 32, 44, 117
 on the march, 111, 115, 117
Speed of artillery, 32
 of cavalry, 26
 of infantry, 43
Squadron, cavalry unit, 26
Squares, formation of, 60, 61, 62
 to be avoided, 61
Stages of attack, 50
Starting, hour of, on the march, 114
Stations in field sketching, 229
Stockades, construction of, 168
 demolition of, 192
Stores, classification of, 163, 173
 collected for passage of river, 157
 for escalade, 187
Storming party, 188, 144, 187
Strategical point, definition of, 2
 village as, 141
Strategy defined, 1

Straw in camp, 8
   contents of rick of, 250
   inspection of, 249
Stream, arrangements for watering at, 8
   measuring with a cap, 219
   reconnaissance of, 79, 241
   to find discharge of, 199
Street, a defile, 151
   patrol passing down a, 76
Strength in artillery, 41
   of advanced guard, 70
   of escorts, 127
   of garrison in village, 142
   of garrison in wood, 148
   of materials, 219
   of outposts, 86
   per mile in positions, 245
Striking tents, 12
Subdivisions, column of, 36
Sunken flying column trench, 15
Supplies of ammunition, 54
   to be tabulated, 242
Support, cavalry in, 29, 78, 109
   extension of, 51
   in a village, 141
   in a wood, 148
   in attack, 51
   of advanced guard, 73
   on outpost duty, 101
Surprise, instances of, 84, 85
   to guard against, 84, 93, 100, 107
Surveying, 221
Suspension bridge, 217
Swamp, report on a, 246
Sword, the arm of cavalry, 25
System of training soldiers, xiv

T.

TACTICAL point, 2
   unit of artillery, 32
   unit of cavalry, 26
   unit of infantry, 43
Tactics defined, 1
Tailing off, evil effects of, 112
Tasks in shelter trenches, 21
Telegraph, destruction of, 195
   poles, to count, 233
   with outposts, 86
Tel-el-Kebir, 121
Tents, circular, capacity of, 10
   drainage for, 12
   parties for pitching, 12
   striking, 12
Theatre of war defined, 2
Timber, to fell, 163, 194
Time marches, 118
   for escalade, 185
   required on the march, 118, 239
   to attack, 102
   to start on the march, 114
Tools for escalade, 187
   for strengthening posts, 173
   issuing, 19
   laying out, 18
Torpedoes, land, 168
Torrens kettles, 15
Touch of the enemy to be maintained, 24, 78, 103
Towns, accommodation in, 240
   reconnaissance of, 79, 240

Trace of field-work, 180
   of gun-pit, 183
   of parallel, 182
   of shelter trenches, 20
Tracks, searching for a ford, 154
Trail bridge, 213
Train, a battalion moving by, 124
   baggage, on the line of march, 112
Training of troops, xiv
Transport company with division, 5
Traverse in field-sketching, 281
Trees, for crossing a stream, 215
   to fell, 163, 194
Trenches, cooking, 15
   shelter, 18, 35, 140, 141
   shelter, filling in, 23
   siege, 182
Trestles, Belgian, 208, 211
   lengthening bridge by, 212
   wooden, 210
Triple arrow kitchen, 16
Tripods for cooking, 17
Troops, disembarking from a train, 125
   in action, space for, 26, 32, 44, 117
   indications of, 247
   moving by rail, 124
   number of, per mile in positions, 245
   on the march, space for, 111, 115, 117
   to conform to the head of the column, 113
Trot, pace at a, 26, 32
Truce, flag of, 95

U.

UNLOADING baggage waggons, 11
Unpacking mules, 196

V.

VALISES, outpost sentries not to take off, 93
Valley-way of a river, 155
Vanguard of advanced guard, 74
Variation of the compass, 230
Vedettes, 96, 109
   how posted, 109
Velocity of rivers, 198
   of sound, 248
Verbal report, 243
Village, accommodation in, 240
   as advanced or detached post, 141
   attack of, 143
   defence of, 138
   intrenching, 140
   patrol entering, 76
   reconnaissance of, 79, 240
Vision, extent of, 247

W.

WAGGONS, American sheet iron, 209
   ammunition, shelter for, 34
   loading baggage, 14
   parking, 131
Wall trench, 15
Walls, destruction of, 191
   treatment of, for defence, 170
War establishment of battalion, 45
   of battery, 40

War establishment of cavalry regiment, 28
Waste weir, 167
Water, allowance of, 199
 arrangement for supply in camp, 8
 convoy by, 133
 depth in fords, 154
 supply, 240
Watering places, 241
Weapons of artillery, 31
 of cavalry, 25
 of infantry, 43
Wheat, yield per acre of, 242
Wheel for measuring distances, 233
 tracks leading to fords, 154
Wickerwork for sentry boxes, 13
 gabions, 175
Wood, attack of, 149
 defence of, 145

Wood, felling, 162
 party in camp, 11
 patrol examining, 76
 piquet in, 100
 reconnaissance of, 79
 roads through, 145
 sentries in, 94
Working parties, 181
 along parallels, extension of, 183
 in attack, 144, 188
 in shelter trenches, extension of, 20
 telling off, 181
Written report, 243

Z.

Zones of fire, three, 3

THE END.

LONDON: MACLURE AND MACDONALD,
Lithographers to the Queen,
97, QUEEN VICTORIA STREET, E.C.

www.ingramcontent.com/pod-product-compliance
Lightning Source LLC
Chambersburg PA
CBHW031131160426
43193CB00008B/106